Environmental Protection

THIRD EDITION

Sarma V. Pisupati

Penn State University

KENDALL/HUNT PUBLISHING COMPANY
4050 Westmark Drive   Dubuque Iowa 52002

Dedication

To my loving family—wife Rama, daughter Sribindu and son Sridhar,
for their help, support and sacrifice of time,

and

To my wonderful students who make me cherish every minute in the classroom

Sarma V. Pisupati

Contents

Preface ix

chapter 1 **Energy and Society** .1
Forms of Energy 2
Measurement of Energy 8
Sources of Energy 10
Power 11
Energy Use of Some Home Appliances 13
Questions 16
Multiple Choice Questions 17
Problems for Practice 19

chapter 2 **Energy Supply and Demand** 21
Global Energy Consumption 22
United States Energy Consumption 29
Growth in the Energy Demand .34
Energy Reserves 35
Questions 38
Multiple Choice Questions 39
Energy Supply and Demand Puzzle 40

chapter 3 **Energy Efficiency** 41
What Is Thermal Energy and How Is It Measured? 44
Questions 52
Multiple Choice Questions 53

chapter 4 **Energy and the Environment** 55
Products of Combustion 56
Health and Environmental Effects of the Primary Pollutants 62
Global and Regional Effects of Secondary Pollutants 65
Impacts of Global Warming 74
Is There a Solution for This Potential Global Warming? 76
Global Warming: Your "Power" in the Environmental Protection 77
Acid Rain 79
Effect of Acid Rain on Health and the Environment 81
Acid Rain: Your "Power" in the Environmental Protection 84

Ozone and Environment 85
Good Ozone or Ozone Layer Destruction 85
International Action 90
Ozone Layer Depletion: Your "Power" in the Environmental Protection . . .91
Ground Level Ozone or "Bad Ozone" Formation and
    Photochemical Smog 91
Ground Level Ozone: Your "Power" in the Environmental Protection 94
Questions 96
Multiple Choice Questions 97

chapter 5   **Appliances** **99**
Appliance Energy Consumption 100
Energy Guide Labels 100
Water Heaters 103
Water Heaters: Your "Power" in the Environmental Protection 113
Refrigerators 114
Refrigerators: Your "Power" in the Environmental Protection 117
Clothes Washers and Dryers 119
Clothes Washers: Your "Power" in the Environmental Protection 121
Dryers 122
Clothes Dryers: Your "Power" in the Environmental Protection 122
Dishwashers 123
Dishwashers: Your "Power" in the Environmental Protection 126
Questions 128
Multiple Choice Questions 129
Problems for Practice 130

chapter 6   **Lighting** **133**
How Is Lighting Measured? 134
How Much Light Is Needed? 135
Types of Lighting 136
Life Cycle Cost Analysis 145
Efficacy of Light Bulbs 147
Questions 152
Multiple Choice Questions 153
Problems for Practice 155
Lighting Puzzle 156

chapter 7   **Home Heating Basics** **157**
Mechanisms of Heat Loss or Transfer 159
Heating Degree Days 162
R-Value 166
Energy Costs 176
Questions 181

Multiple Choice Questions 182
Problems for Practice 184
Home Heating Basics Puzzle 189

chapter 8   **Home Heating Systems** **191**
Central Ducted Air Systems 193
Radiant Heating Systems 195
Direct or In-Situ Heating Systems 201
Cooling and Heating/Cooling Systems 205
Heat Pumps 206
Ground Source (Geothermal) Heat Pumps 210
Factors Affecting the Type of GHP Loop 215
Solar Energy for Home Heating 219
Home Heating: Your "Power" in the Environmental Protection 223
Questions 225
Multiple Choice Questions 226
Problems for Practice 228

chapter 9   **Home Cooling** **229**
How Do We Measure Humidity? 230
How an Air Conditioner Works 232
Types of Air Conditioners 233
Air Conditioner Efficiency 235
Home Cooling: Your "Power" in the Environmental Protection 244
Questions 246
Multiple Choice Questions 247
Problems for Practice 248

chapter 10   **Windows** **251**
Factors in Window Selection 255
Advances in Window Technologies 257
Smart Windows 265
Questions 269
Multiple Choice Questions 270
Problems for Practice 271
Windows Puzzle 272

Explanation of Selected Terms Related to Energy and
    Environmental Protection 273
Answers to Multiple Choice Questions 281
Answers to Numerical Problems 283
Useful Conversion Factors 285
Useful Formulae 287
Solutions to Selected Problems 291

Preface

Energy is a vital component of modern society. Much of the general population believes that the energy sources we depend on are perpetual. While people believe that energy use is the culprit for environmental damage, they are not aware of the methods and principles by which energy conversion devices operate. This introductory book provides necessary knowledge and information about the main operating principles of devices/appliances that are in common use, and information with which one can make the most energy-efficient and economical choice. These devices are day-to-day appliances such as refrigerators, washers and dryers, ovens, etc. and home heating or cooling and transportation choices. The objective of this book is to expose readers to energy efficiency in day-to-day life to save money and energy and thereby protect the environment. This education is very important in order for all students to become environmentally responsible individuals of this Global Village.

This book is compiled from several governmental and other informational sources and is designed for undergraduate students with no science or engineering prerequisites. It consists of ten chapters. The first three chapters focus on energy basics—on how much and how efficiently we use energy—and the fourth chapter discusses the relationship between energy use and environmental impact. Chapters five through seven focus on home heating and cooling requirements and efficiency improvements that can save energy and money and will protect the environment. The last three chapters provide basic insight into energy-efficient windows, lighting, and home appliances.

This task of compiling the information and turning it into a book that students, with little background in science, can understand is not an easy task without help from others. I am grateful to all of them. My colleague Dr. Jonathan Mathews reviewed the manuscript and provided valuable suggestions. Amanda Rauer and Rebecca Entler provided comments from a student perspective. My teaching assistants over the years also helped me develop some of the problem sets. Prasanna Chidambaram and Prabhat Naredi have provided valuable input. Nick Smerker, Mark DeLuca, and Kelly Henry created some of the quality graphics that I used in the book.

I hope readers enjoy this book and recognize their "power" in protecting the environment!

Sarma V. Pisupati

Energy and Society

goals

- ☞ *To define energy*

- ☞ *To articulate fundamental forms of energy*

- ☞ *To know different units of energy*

- ☞ *To define and distinguish differences between energy and power*

- ☞ *To classify energy sources*

*E*nergy is the life blood of any society, without which the modern life would come to a standstill. From the moment we wake up in the morning until the time we hit the bed, we use energy for almost everything we do—even while sleeping comfortably at night (using heating or air conditioning); making a hot cup of coffee in the morning; taking a shower (using water heating); preparing breakfast, lunch, or dinner (cooking); storing food in the refrigerator (appliances); going to and coming from work (transportation); communicating at work, preparing presentations, archiving information (computers, emails, faxes, overnight mail); watching TV or going to a movie (recreation); traveling on vacation and for business (transportation); making the goods that we need (industries); and bringing the goods to the consumer (commercial), and so on. Therefore, it is expected that the quality of life a society enjoys depends on the energy consumption of that society as shown in Figure 1.1.

"Energy is the property of matter that can be converted to work, heat, or radiation"

When energy is so important, the questions that come to our mind are: What is energy? How do we measure it? Where is it coming from? Do we have enough? And what is the economic and environmental impact of energy use? To answer these questions, we need to understand the term *energy*. From the examples mentioned above we can say that energy is that which has the ability to move things or to do work. It is also that which can produce heat, even if it does not move anything. Energy can also be that which can be converted to light (more accurately called radiation).

Forms of Energy

Energy exists in a number of different forms, all of which give the ability to an object or system to do work on another object or system. There are six different

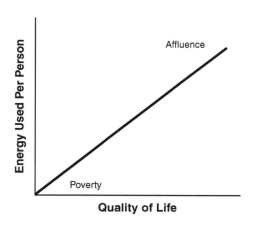

chapter 1   *Energy and Society*

Figure 1.1   *Relationship of energy use to the quality of life.*

basic forms of energy that we use in our day-to-day life: mechanical, chemical, heat, electrical, radiant, and nuclear.

1. Mechanical Energy

a. Kinetic Energy

Kinetic energy is the energy that a body possesses by virtue of its motion. Consider a baseball flying through the air or a car moving on the road. The ball and the car are said to have **kinetic energy** by virtue of the fact that they both are in motion. You can see that the baseball has energy because it can do "work" on an object on the ground if it collides with it (either by pushing on it and/or damaging it during the collision). Similarly, a moving car is doing work against gravitational force.

> ❝Kinetic energy is the energy that a body possesses by virtue of its motion❞

Practical Application: In the United States we use about 27 percent of our total energy for the transportation or movement of people and goods.

b. Potential Energy

Potential energy is the energy that a body possesses by virtue of its position relative to a reference point. Consider a book sitting on a shelf in the library (resting at a height). The book is said to have **potential energy** (stored energy). If it is nudged off, gravity will accelerate the movement of the book, giving it kinetic energy (due to motion). The Earth's gravity is necessary to create kinetic energy in the book in its resting state. This gravity in turn depends on the "Earth-book system" called Gravitational Potential Energy. This potential energy is converted into kinetic energy as the book falls. Another example is a mechanical spring that is wound. As the spring is released the energy is released.

> ❝Potential energy is the energy that a body possesses by virtue of its position relative to a reference point❞

2. Chemical Energy

Chemical energy is the energy that is locked in the bonds of a molecule that hold the atoms together, as shown in Figure 1.2. When these bonds are broken and new bonds are formed, energy is released. The food that we eat is converted into glucose in the blood. The glucose (blood sugar) in your body is said to have **chemical energy** because glucose combines with the oxygen we breathe and thereby releases energy. Our muscles use this chemical energy to generate mechanical energy and heat. Therefore, chemical energy is really the energy locked in the bonds of molecules in the

> ❝Chemical energy is the energy that is locked in the bonds of a molecule that hold the atoms together❞

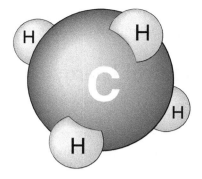

Figure 1.2

Methane molecule with carbon and hydrogen atoms holding together.
Sᴏᴜʀᴄᴇ: *http://www.energy.gov.ab.ca/gmd/images/Co*

form of microscopic potential energy (stored energy), which exists because of the electric and magnetic forces of attraction exerted between the different atoms of the molecule. These parts get rearranged in chemical reactions, releasing or adding to this potential energy.

Practical Application: Approximately 85 percent of the energy used in the United States comes from fossil fuels such as coal, oil, and natural gas. All these fuels store energy in the form of chemical energy. When burnt, these fuels release energy in the form of heat or thermal energy.

3. Thermal or Heat Energy

Heat energy is the kinetic energy of the molecules. Consider a hot cup of coffee. The coffee is said to possess **thermal energy**, or **heat energy**, which is really the collective, minute kinetic and potential energies of the molecules in the coffee. (The molecules have kinetic energy because they are moving and vibrating, and they have potential energy due to their mutual attraction for one another—much the same way as the "Earth-book system" does.) Thermal energy can be measured by temperature. At higher temperatures, the molecules move around and/or vibrate faster, increasing the kinetic and potential energy of the molecules.

Practical Application: When fuels (chemical energy) are burnt, the energy converts to thermal energy (heat), which is then converted to kinetic energy (motion) in an automobile or into electricity in the case of a power plant.

4. Electrical Energy

All matter is made up of atoms, and atoms are made up of smaller particles, called protons (that are positively charged), neutrons (that are neutrally charged), and electrons (that are negatively charged). Electrons orbit around the nucleus, which contains protons and neutrons, in the same way that the planets orbit the sun.

Certain metals have electrons that are only loosely attached to their nuclei. They can easily be made to move from one atom to another if an electric field is applied to them. When those electrons move among the atoms of the matter, **electricity** is produced. Although electricity is seldom used directly, it is one of the most useful and versatile forms of energy. For instance, electricity put into a toaster gets converted to heat, put into a stereo gets converted to sound, put into an electric bulb gets converted to light (and heat), and put into a motor gets converted to motion or (mechanical energy). Due to its versatility, electricity is in high demand.

Practical Application: In the United States about 40 percent of the total primary energy used is converted into electricity for various uses.

5. Radiation

Consider the energy transmitted to the Earth from the sun. This energy is called **radiation**, which includes visible light that can be seen by the naked eye, infrared radiation, and ultraviolet radiation (UV) that cannot be seen by the naked eye. Radiation can also include long-wave radiation such as TV waves and radio waves and very short waves such as X-rays and gamma rays. Hence, the entire spectrum of radiation is also called "electro-magnetic radiation" as shown in Figure 1.3. Electromagnetic radiation oscillates (from side to side) coupled with electric and magnetic fields that travel freely through space.

Light can also be considered as little packets of energy called photons (that are actually particles, rather than waves). The word "photon" is derived from

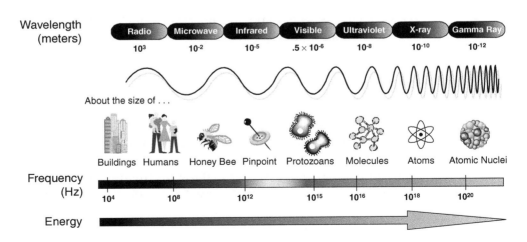

Figure 1.3

The electromagnetic spectrum.

the word "photo," which means "light." When energy is supplied to atoms, electrons in an atom jump to higher and excited levels. Photons are produced when these excited electrons fall from the higher (excited) energy level back to lower (ground state) energy levels. Photons are also generated when a charged particle, such as an electron or proton, is accelerated as, for example, happens in a radio transmitter antenna that generates radio waves.

In general, the weaker the energy, the longer the wavelength and lower the frequency, and vice versa. The reason that sunlight can hurt your skin or your eyes is because it contains "ultraviolet light," which consists of high-energy photons. These photons have a short wavelength and a high frequency, and enough energy is packed in each photon to cause physical damage to your skin if they get past the outer layer of skin or the lens in your eye. Radio waves, and the radiant heat you feel at a distance from a campfire, for example, are also forms of electro-magnetic radiation, or light, except that they consist of low-energy photons (long wavelength but high frequencies) that your eyes can't perceive. A great discovery of the nineteenth century was that radio waves, X-rays, and gamma-rays, are all various forms of light, and that light consists of electromagnetic waves.

Practical Application: About 20 percent of the electricity produced in the United States is used for lighting purposes to produce visible light.

6. Nuclear Energy

Nuclear energy is the energy generated when there are changes in the nuclei of the atoms. It is interesting to note that these nuclear reactions that occur in the sun are responsible for solar energy. These reactions also occur in nuclear reactors and in the interior of the Earth. In the sun, hydrogen nuclei fuse (combine) together to make helium nuclei, in a process called fusion, whereby energy is released. On the other hand, in the case of a nuclear reactor, or in the interior of the Earth, uranium nuclei (and certain other heavy elements in the Earth's interior) split into smaller particles, releasing huge amounts of energy. This process is called fission. Hydrogen which is the source for nuclear fusion is available in water. Almost 70 percent of the planet's surface is covered with water and, therefore, fusion can be classified as renewable energy. On the other hand, nuclear fission is a non-renewable energy source because the source (uranium) is finite. The energy released by fission and fusion is not just a product of the potential energy released by rearranging the electrons as in chemical reactions. In fusion or fission, some of the matter making up the nuclei is actually converted to energy. The amount of energy released is described by the famous equation,

$$E = mc^2 \qquad (1.1)$$

where E = Energy, m = Mass destroyed, and c = Speed of light.

Fundamental Forms of Energy

- ☞ Mechanical
- ☞ Chemical
- ☞ Thermal or Heat
- ☞ Radiation
- ☞ Electrical

Equation 1.1 explains that when the matter is destroyed, the energy stored is released. It is important to know that we do not use fusion for commercial energy production.

We know from our experience that by pumping gasoline into our cars and with the help of electrical energy from the car battery we obtain mechanical energy (motion of the vehicle). We also know that when we buy electrical energy and feed it into a TV, electrical energy gets converted to light and sound. Similarly, an electric bulb converts electrical energy into visible light and small amounts of heat. Figure 1.4 shows examples of some such conversions. From these examples we also learn that energy can be converted from one form to another.

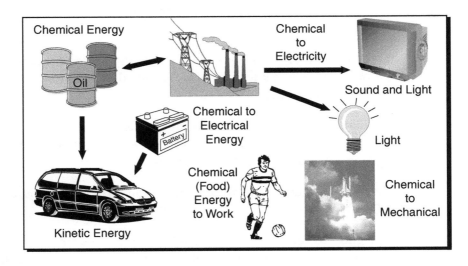

| Figure 1.4 |

Examples of day-to-day transformations.

Measurement of Energy

Energy is measured in various units by various industries or countries just like the value of goods is expressed in various currencies such as dollars in the United States, yen in Japan, and pounds in Britain

Units

British thermal unit (BTU): A unit of energy equal to the amount of energy needed to raise the temperature of one pound of water by one degree Fahrenheit. To easily visualize this, think of it as the approximate amount of energy contained in the tip of a matchstick. This unit is used mostly by the heating and cooling industry.

Calorie (also small calorie): The amount of energy needed to raise the temperature of one gram of water by one degree Celsius. One calorie equals 0.003969 British thermal units (BTU). Note that this calorie is written with a lower case "c."

Units of Energy

- ☛ Joule
- ☛ BTU
- ☛ calorie
- ☛ Calorie or Food calorie
- ☛ kWh
- ☛ ft lb

Food calorie: The food calorie is often used when measuring the energy content of food. [Cal, kcal; also Calorie (written with a capital "C"). A kilocalorie is equal to 1,000 calories.] A unit of food energy is equal to one kilocalorie.

Joule (J): A unit of energy used by scientists and engineers. It is a smaller quantity of energy than a calorie and very much smaller than a BTU. One joule equals 0.2388 calories or 0.0009481 BTU.

Kilocalorie: [Cal, kcal; also Calorie (written with a capital "C"). Food calorie, Large calorie.] The amount of energy needed to raise the temperature of one kilogram of water one degree Celsius. It is equal to 1,000 calories, 4,187 joules, or 3.969 BTUs.

Kilowatt-hour (kWh): An amount of energy from the steady production or consumption of one kilowatt of power for a period of one hour. A unit of energy equal to 3,413 BTUs or 3,600,000 joules.

Therm: A unit describing the energy contained in natural gas. One therm equals 100,000 BTUs.

Figure 1.5 shows the energy used by various countries or processes on a logarithmic scale. A logarithmic scale is useful when a wide range of numbers of a variable needs to be plotted. For example, on a normal graph if each cm is rep-

Energy Scale

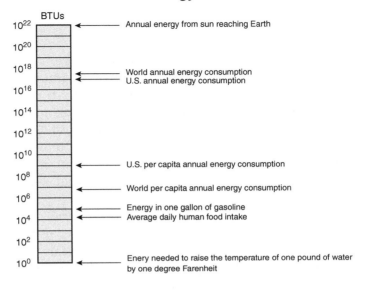

Figure 1.5 *Energy scale.*

resented as one BTU, then point 1 is the origin, point 2 is one BTU, point 3 is two BTUs, and point 4 is three BTUs. On a logarithmic scale, point 1 is the origin, but point 2 is 10^2 or 100 BTUs, and point 3 is 10^3 or 1,000 BTUs. So with three points or three cm we can represent three BTUs, but on a logarithmic scale with the same three cm we represent 10^3 or 1,000 BTUs. So with n points on a normal scale we can represent n BTUs, but on a logarithmic scale we can represent $10n$ BTUs. There are several quantities that are expressed on a logarithmic scale. One such quantity is acidity as measured by pH as we will see later in Chapter 4.

A BTU is an amount of energy that can raise the temperature of one pound of water by one degree Fahrenheit. It is also approximately the amount of energy contained in the tip of a matchstick. The average food intake of a person is about 10,000 BTUs a day, and each gallon of gasoline has about 125,000 BTUs or 12.5 times more than the average daily human calorie intake.

On the average, every person in the United States used about 342,678,430 (342.7×10^6) BTUs in the year 2004 compared to an average consumption of 70,055,000 BTUs per person in the world. The energy consumption of the United States in 2004 was 100,351,000,000,000,000 (100.3×10^{15}) BTUs, and that of the world is 446,440,000,000,000,000 (446.44×10^{15}) BTUs.

It is interesting to note that the Earth received approximately 24,000 times more energy than what the entire world used in 2004. In that case, one might ask "why should anybody worry about energy shortage?" The real issue is not whether there is enough energy available, but rather the form in which it is available and whether it can be easily converted to the form that we need.

Sources of Energy

The sources of energy can be divided into two groups: renewable (an energy source that can be replenished over and over again) and non-renewable (an energy source that we are using up and that cannot be produced in a short period of time). The five renewable sources that are commonly used include hydropower (water), solar, wind, geothermal, and biomass. Renewable energy sources can be converted to electricity and heat. On the other hand, geothermal energy from within the Earth, wind energy from uneven heating of the Earth's surface, biomass from plants, and hydropower from water are some of the common renewable energy sources. Classification of sources of energy is shown in Figure 1.6.

We get most of our energy from non-renewable energy sources, which include the fossil fuels: oil, natural gas, and coal. They're called fossil fuels because they were formed over millions and millions of years by the action of heat from the Earth's core and pressure from rock and soil on the remains (or "fossils") of dead plants and animals (microscopic). Another non-renewable energy source is the element uranium, whose atoms when split (through a process called nuclear fission) generate heat and ultimately electricity.

The fossil fuels that formed over millions of years are not distributed uniformly over the Earth's surface. Depending on the climatic conditions that prevailed millions of years ago, certain parts of the land masses were favorable for the growth of the organic matter. Over the geological ages, these land masses

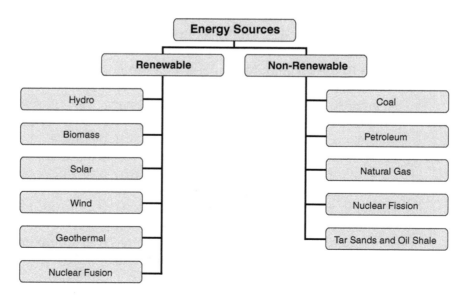

Figure 1.6   *Classification of sources of energy.*

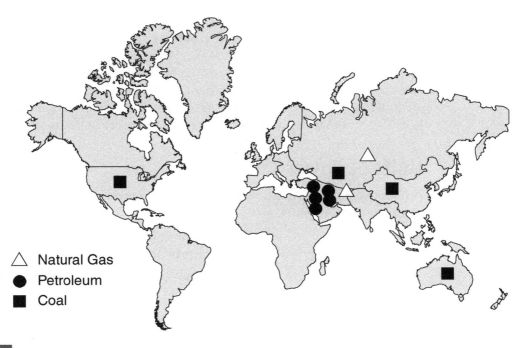

△ Natural Gas
● Petroleum
■ Coal

Figure 1.7   *Distribution of fossil fuels.*

moved; hence, certain regions are richer in fossil fuels than others. As shown in Figure 1.7, the United States, China, and Australia are rich in coal deposits, and Middle Eastern countries and Russia are rich in oil reserves and natural gas. Therefore, imports and exports of fossil fuels and international politics play an important role in energy supply and security.

Power

Power is the rate at which energy is used. Some units of measure of power include horsepower and watts. One way to help distinguish between work and power is to think of two people each walking four miles. One person walks faster than the other. They both walk the same distance (same amount of work is involved), one just walks at a faster rate (the faster one was a power walker!).

$$Power = \frac{Energy}{Time} \qquad (1.2)$$

or

$$Energy = Power \times Duration\ of\ Usage\ (Time)$$

The units of power will always have energy units divided by time or per unit time. BTUs and Cal are energy units, but BTUs/h and Cal/sec are units of power. Energy is like the money you make, and power is like the rate at which you make money. For example, if you made $50, you made $50. You did not say how fast you made that money. Whereas when you say that you make $50 an hour, you are indicating how fast you make money but not how much money you make. Why? The money you make depends on the number of hours you work.

Units of Power

Horsepower (hp): A unit of power. One horsepower equals 550 foot-pounds per second or 746 watts.

Watt (W): A unit of power. One watt equals the production or use of one joule of energy per second (1 J/s).

Kilowatt (kW): A unit of power equal to 1,000 watts. One MW is equal to 1,000 kW or one million W.

$$Power = \frac{Energy}{Time}$$

Energy consumption per day = Power consumption × Hours used per day

Prefixes are sometimes used to show either small or large quantities. Table 1.1 shows some commonly used prefixes

Table 1.1	*Common Prefixes*	
pico-	p	10^{-12}
nano-	n	10^{-9}
micro-	m	10^{-6}
milli-	m	10^{-3}
centi-	c	10^{-2}
deci-	d	10^{-1}
		1
deka-	D	10^{1}
hecto-	h	10^{2}
kilo-	k or K	10^{3}
mega-	M	10^{6}
giga-	G	10^{9}
tera-	T	10^{12}
peta-	P	10^{15}
exa-	E	10^{18}

Energy Use of Some Home Appliances

In order to calculate how much electricity is consumed by your home appliances and to compare it with your electricity bill later we need to know the power consumption of each of those appliances and the duration of use. Table 1.2 provides the power consumption (Wattage) of some typical home appliances.

$$Energy\ consumption\ per\ day\ =\ Power\ consumption \times Hours\ used\ per\ day$$

Recall that 1 kilowatt hour (kWh) = 1,000 Watts hours. Therefore,

$$\frac{\text{Daily kilowatt-hour (kWh)}}{\text{consumption}} = \frac{\text{Energy consumption per day (Wh)}}{1,000\ \text{Wh}}$$

Multiply this by the number of days you use the appliance during the year for the annual consumption. You can then calculate the annual cost to run an appliance by multiplying the kWh per year by your local utility's rate per kWh consumed.

Units of Power

- ☞ BTUs/h
- ☞ J/s (Watt)
- ☞ HP

EXAMPLES

Ceiling Fan:

Assuming that we use the ceiling fan for only four hours a day and for 120 days in a year:

$$\frac{\text{Annual energy}}{\text{consumption}} = \frac{200\ \cancel{Watts}}{1000\ \cancel{Wh}}\ \frac{1\ kWh}{}\times \frac{4\ \cancel{h}}{\cancel{day}} \times \frac{120\ \cancel{days}}{year} = 96\ kWh$$

$$Annual\ cost\ =\ Annual\ energy\ consumption\ (kWh) \times Price\ per\ kWh$$

$$Cost\ =\ 96\ \cancel{kWh} \times \frac{\$0.0845}{\cancel{kWh}} = \$8.12$$

Personal Computer and Monitor:

Total wattage = 120 + 150 Watts = 270 Watts

Again, assuming that we use the computer for only four hours a day but all days in a year:

$$\frac{\text{Annual energy}}{\text{consumption}} = \frac{270\ \cancel{Watts}}{1000\ \cancel{Wh}}\ \frac{1\ kWh}{}\times \frac{4\ \cancel{h}}{\cancel{day}} \times \frac{365\ \cancel{days}}{year} = 394.2\ kWh$$

$$Cost\ =\ 394.2\ \cancel{kWh} \times \frac{\$0.0845}{\cancel{kWh}} = \$33.3 = \$33.3\ per\ year$$

Table 1.2

Typical Range of Power Consumption (Wattage) of Some of the Commonly Used Appliances

APPLIANCES	WATTAGE (RANGE)
Aquarium	50–1,210 Watts
Clock radio	10
Coffee maker	900–1,200
Clothes washer	350–500
Clothes dryer	1,800–5,000
Dishwasher	1,200–2,400 (using the drying feature greatly increases energy consumption)
Dehumidifier	785
Electric blanket—Single/Double	60/100
Fans	
Ceiling	65–175
Window	55–250
Furnace	750
Whole house	240–750
Hair dryer	1,200–1,875
Heater (portable)	750–1,500
Clothes iron	1,000–1,800
Microwave oven	750–1,100
Personal computer	
CPU—awake/asleep	120/30 or less
Monitor—awake/asleep	150/30 or less
Laptop	50
Radio (stereo)	70–400
Refrigerator (frost-free, 16 cubic feet)	725
Televisions (color)	
19"	65–110
27"	113
36"	133
53"–61" Projection	170
Flat screen	120
Toaster	800–1,400
Toaster oven	1,225
VCR/DVD	17–21/20–25
Vacuum cleaner	1,000–1,440
Water heater (40 gallon)	4,500–5,500
Water pump (deep well)	250–1,100

You can usually find the wattage of most appliances stamped on the bottom or back of the appliance, or on its "nameplate." The wattage listed is the maximum power drawn by the appliance. Since many appliances have a range of settings (for example, the volume on a radio), the actual amount of power consumed depends on the setting used at any one time.

Refrigerators, although turned "on" all the time, actually cycle on and off at a rate that depends on a number of factors. These factors include how well the refrigerator is insulated, room temperature, freezer temperature, how often the door is opened, if the coils are clean, if it is defrosted regularly, and the condition of the door seals. To get an approximate figure for the number of hours that a refrigerator actually operates at its maximum wattage, divide the total time the refrigerator is plugged in by three.

If the wattage is not listed on the appliance, you can still estimate it by finding the current draw (in amperes) and multiplying that by the voltage used by the appliance. Most appliances in the United States use 120 volts. Larger appliances, such as clothes dryers and electric cook tops, use 240 volts. The amperes might be stamped on the unit in place of the wattage. If not, find a clamp-on ammeter—an electrician's tool that clamps around one of the two wires on the appliance—to measure the current flowing through it. You can obtain this type of ammeter in stores that sell electrical and electronic equipment. Take a reading while the device is running; this is the actual amount of current being used at that instant.

Also note that many appliances continue to draw a small amount of power when they are switched "off." These "phantom loads" occur in most appliances that use electricity, such as VCRs, televisions, stereos, computers, and kitchen appliances. Most phantom loads will increase the appliance's energy consumption a few watts per hour. These loads can be avoided by unplugging the appliance or using a power strip and using the switch on the power strip to cut all power to the appliance.

Sources

http://www.eia.doe.gov/kids/whatsenergy.html
http://www.eere.energy.gov/consumerinfo/factsheets/ec7.html

questions

1. List the six fundamental forms of energy.

2. What is the difference between energy and power?

3. List four devices that you use at home and write the input and output forms of energy to and from those devices.

4. Explain the difference between renewable and non-renewable energy sources.

5. Most of the day-to-day devices that we use convert energy from one form to another. The table below lists some day-to-day devices. Identify the fundamental forms of energy that are put in and the energy output that comes out.

Input and Output Forms of Energy in Some Devices/Sources

INPUT FORM OF ENERGY	DEVICE/SOURCE	OUTPUT FORM OF ENERGY
	Lawn Mower	
	Computer	
	Sun	
	Tree	
	Gas Furnace	
	Hair Dryer	

multiple choice questions

1. What is the unit of power?
   a. Watt hour
   b. Joule
   c. BTU/h
   d. Quad

2. Energy absorbed by shock absorbers in a car is stored in the shocks as kinetic energy.
   a. True
   b. False

3. One calorie is defined as the amount of energy that is required to raise the temperature of
   a. 1 lb of water through 1°F
   b. 1 g of water through 1°C
   c. 1 lb of water through 1°C
   d. 1 gal of water through 1°C

4. Energy is defined as a property of matter that can be converted to
   a. Work
   b. Heat
   c. Radiation
   d. All the above
   e. None of the above

5. What is the unit of power?
   a. Volts
   b. Joule/s
   c. BTU
   d. kWh

6. Infrared waves have more energy than ultraviolet waves
   a. True
   b. False

7. Energy and work have the same units.
   a. True
   b. False

8. How much energy is in one BTU?
   a. Enough to heat one gram of water one degree Celsius
   b. Enough to heat one gallon of water one degree Fahrenheit
   c. Enough to heat your room to a comfortable temperature
   d. Enough to heat one pound of water one degree Fahrenheit

9. A 100-watt incandescent light bulb is operated for 120 hours, and a 15-watt fluorescent light bulb is operated for the same period of time. At 10 cents per kWh, what is the cost savings of the fluorescent bulb?
   a. $1.02   c. $0.01
   b. $0.10   d. $0.50

10. Energy can be converted from one form to another.
    a. True
    b. False

11. Which form of energy is present in a football traveling above the ground in the air?
    a. Chemical
    b. Heat
    c. Mechanical
    d. Electrical

12. kWh is the unit of energy.
    a. True
    b. False

13. Most of the energy that the United States needs today is coming from
    a. Petroleum
    b. Solar
    c. Nuclear
    d. Biomass

14. Glucose in your body is an example of
    a. Mechanical energy
    b. Electrical energy
    c. Heat energy
    d. Chemical energy

15. Coal is a renewable energy source.
    a. True
    b. False

16. Geothermal energy is a renewable energy source.
    a. True
    b. False

17. Solar energy is classified as a renewable energy source.
    a. True
    b. False

18. Natural gas is a non-renewable energy source.
    a. True
    b. False

19. The form of energy in the pizza we eat is
    a. Mechanical
    b. Food
    c. Chemical
    d. Thermal

20. A thermal calorie is a unit of energy.
    a. True
    b. False

21. Watt is the unit of energy.
    a. True
    b. False

22. Power multiplied by time gives energy.
    a. True
    b. False

23. Therm is generally used to quantify the energy contained in natural gas.
    a. True
    b. False

24. Most of the energy demand in the U.S. is met by renewable energy sources.
    a. True
    b. False

25. Most of the energy for the U.S. is coming from nuclear energy.
    a. True
    b. False

26. A BTU has more energy than a calorie.
    a. True
    b. False

problems for practice

1. Michael weighs 180 lbs and burns 600 Cal/h when jogging at 6 miles/h and burns 300 Cal/h when walking at 3 miles/h. On a typical day Michael jogs for 15 minutes and walks for 45 minutes. How many Calories (energy spent) were burnt during this exercise?

2. A water heater consumes 3,000 Watts of power and operates for about 6 hours a day. Calculate the annual operating cost of the water heater in your region. Assume a price of $0.10/ kWh of electrical energy.

3. Michael uses his toaster to make toast for his breakfast. The toaster is rated for 1,200 W, and he uses it for 15 minutes a day. How much energy is used in a month by this toaster?

4. A refrigerator consumes 150 Watts of power. Calculate the annual operating cost of a refrigerator in your region. Assume that the refrigerator operates for 4,000 hours a year and that the electricity costs $0.08 for a kWh.

5. A typical computer consumes 150 Watts of power. Jackie turns off her computer after using it for 4 hours every day. Her roommate Stacie leaves the computer on after finishing her work (same 4 hours) with an "away message" on for all 24 hours. By the end of the month what would be the energy consumption for each of these roommates? At 8.5 cents per kWh what would be the difference in cost to operate the computers?

6. A 36-inch TV consumes 200 Watts of power. Calculate the annual operating cost of the TV if it is operated 4 hours a day. Assume that electricity costs 10 cents/kWh.

7. A 100-watt incandescent light bulb is operated for 6 hours a day for one full year, and a 25-watt fluorescent light bulb is operated for the same period of time. At 10 cents per kWh, what is the cost savings of the fluorescent bulb?

8. A water heater consumes 4,000 Watts of power and operates for about 10 hours a day. Calculate the annual operating cost of the water heater in your region. Assume a typical price for kWh of electrical energy.

Activity

A typical household uses the appliances rated as follows. Estimate the duration for which you use the appliance and calculate the total energy use (kWh) for the month of September and the cost at $0.065/kwh.

APPLIANCE	ESTIMATED POWER RATING (WATTS)	ESTIMATED DURATION OF USE/DAY (H)	ENERGY CONSUMPTION/DAY
Refrigerator*	725		
Dishwasher	1500		
Clothes iron	1500		
Television	100		
Microwave oven	800		
Clothes washer	400		
Clothes dryer	2000		
Coffee maker	1000		
PC	100		
Water heater*	4500		
Vacuum cleaner	1000		
Toaster	800		
Freezer*	1000		
AC	750		
Lights	75 each		

*Assume that they run for 4 hours a day.

Total kWh in the month of September:

Estimated cost of electricity:

Energy Supply and Demand

goals

- To analyze global and national energy consumption patterns

- To define energy intensity

- To understand growth in energy consumption and calculate future energy demand

- To obtain knowledge about the energy reserves of the United States and the world, and to appreciate their estimated lifetimes

Global Energy Consumption

The world's total primary energy consumption in 2004 was 446.44 quadrillion BTUs. Figure 2.1 shows various regions and their total primary energy consumption in 2004.

Energy consumption is high for North America. The amount of energy used depends on the economic prosperity of the nation and the population of the country. The North American region includes Canada, the United States, and Mexico. The Far East and Oceania include developed nations such as Japan and Australia, and densely populated developing nations such as China and India. This figure, however, does not reflect the energy intensity. Energy intensity is the ratio of energy consumption to a measure of the demand for services (e.g., number of buildings, total floor space, floor space-hours, number of employees, or constant dollar value of Gross Domestic Product for services). Figure 2.2 shows the total primary energy consumption per capita (energy intensity). The productivity of a country is measured by the total value (dollars) of goods and services produced by its people; this is called the Gross Domestic Product (GDP). Therefore, the average value of goods and services produced by each person (GDP per capita) of a country is an indicator of the productivity of a nation.

Total Primary Energy Use = 446.44 Quads

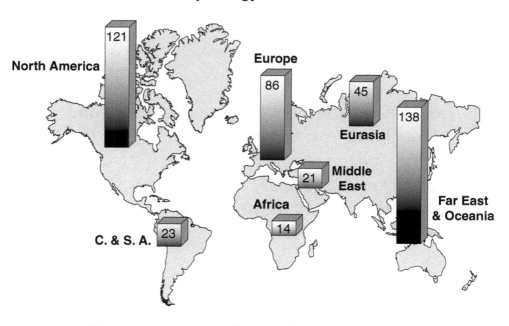

| Figure 2.1 | *Total primary energy use of different regions in the world (2004).* |

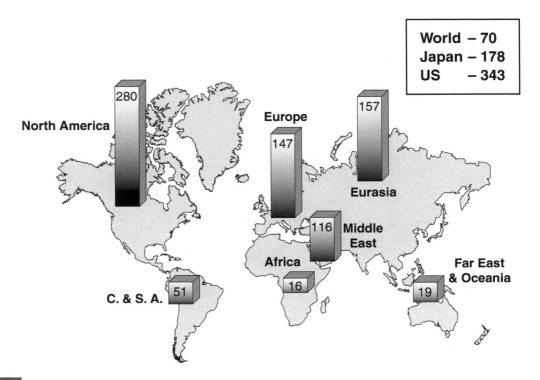

Figure 2.2 *Total primary energy consumption per capita (MMBTU/person) in 2004.*

In the industrialized countries, history shows the link between energy consumption and economic growth to be a relatively weak one, with growth in energy demand lagging behind economic growth. In the developing countries, the two have been more closely correlated, with energy demand growing parallel with economic expansion. Figure 2.3 shows the correlation between per capita energy consumption and per capita GDP (productivity) for selected countries. In general, as the GDP per person of any country increases, the amount of energy required is also expected to increase. The correlation is much stronger for developing nations, but for developed nations the correlation is weak. For example, France and Singapore with similar GDP per capita (31,100 and 31,400 dollars per person, respectively) have significant differences in energy consumption per capita (225.2 and 444.6 MMBTUs per person, respectively). Figure 2.4 shows the energy intensity of several countries in 2004. A nation's GDP at purchasing power parity (PPP) exchange rates is the sum value of all goods and services produced in the country valued at prices prevailing in the United States. This is the measure most economists prefer when looking at per-capita welfare and when comparing living conditions or use of resources across countries.

There are obviously differences in the industrial, transportation, commercial, and residential energy efficiencies. However, differences in the climatic and geographical area of a country can also account for some differences. Differences

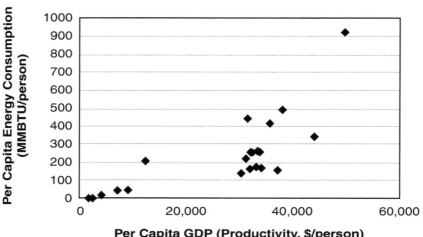

Figure 2.3   *Energy consumption as a function of productivity of selected countries.*

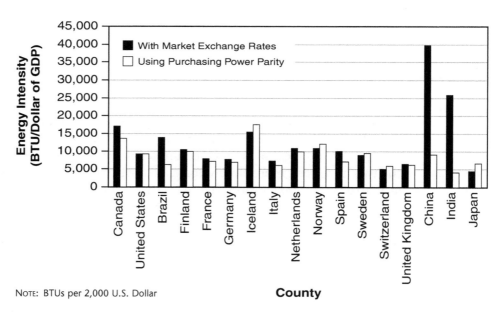

NOTE: BTUs per 2,000 U.S. Dollar

Figure 2.4   *Energy intensity of selected countries.*

in the lifestyles (use of more gas-guzzling cars and SUVs, and bigger houses) and the nature of the products (steel vs. tulips) produced by the nations' industries are also important factors.

The United States Department of Energy projects strong growth for worldwide energy demand over the 24-year projection period from 2004 to 2030. Total world consumption of marketed energy is projected to increase from 447 quadrillion

BTU in 2004 to 559 quadrillion BTU in 2015 and then to 702 quadrillion BTU in 2030—a 57-percent increase over the projection period. The fastest growth is projected for the nations of developing Asia, including China and India, where robust economic growth accompanies the increase in energy consumption over the forecast period. Gross domestic product (GDP) in developing Asia is expected to expand at an average annual rate of 5.8 percent, compared with 4.1 percent per year for the world as a whole. With such strong growth in GDP, demand for energy in developing Asia doubles over the forecast, accounting for 50 percent of the total projected increment in world energy consumption and 77 percent of the increment for the developing world alone.

Current and Future Energy Sources

The world's energy supply sources for the year 2004, and the projected supply for the year 2030, are shown in Figures 2.5 and 2.6, respectively.

Oil Demand

☞ 2004—168 Quads

☞ 2030—239 Quads

42% increase

Over the past several decades, oil has been the world's foremost source of primary energy consumption, and it is expected to remain in that position throughout the period 2004 to 2030. Liquids (primarily oil and other petroleum products) are expected to continue to provide the largest share of world energy consumption over the projection period, but it is also expected that their share will fall from 38 percent in 2004 to 34 percent in 2030 largely because rising world oil prices will dampen the demand for liquids after 2015. Worldwide liquids consumption is projected to increase from 83 million barrels per day in 2004 to 97 million barrels per day in 2015 and 118 million

2004 Total Energy Consumption = 446.44 Quads

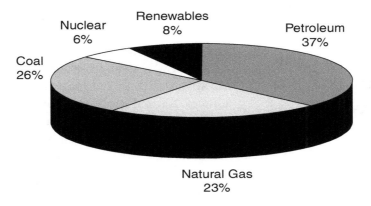

| Figure 2.5 | *World energy consumption by source (2004).* |

DATA SOURCE: *U.S. Energy Information Administration, International Energy Outlook 2007.*

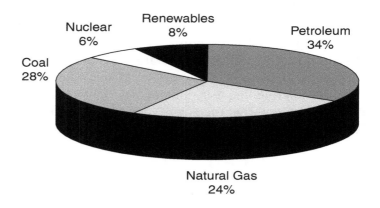

2030 Total Energy Consumption = 702.0 Quads

Nuclear 6%
Renewables 8%
Petroleum 34%
Coal 28%
Natural Gas 24%

Figure 2.6   *Projected world energy consumption by source in 2030.*
DATA SOURCE: *U.S. Energy Information Administration, International Energy Outlook 2007.*

barrels per day in 2030. Petroleum products remain the most important fuels for transportation because there are few alternatives that can be expected to compete widely with petroleum-based liquids; however, the role of oil outside the transportation sector continues to be eroded because of high world oil prices in most regions of the world.

As a result, oil is projected to retain its predominance in the global energy mix and to meet 34 percent of the total primary energy consumption in 2030.

Rising world oil prices increase the demand for natural gas, as it is used to displace the use of liquids in the industrial and electric power sectors in many parts of the world. Industrial uses throughout the world are projected to make up 43 percent of total natural gas use in 2030. It is the world's fastest-growing energy source for electricity generation, leading to an increase in the electric power sector share of total natural gas use worldwide, from 31 percent in 2004 to 36 percent in 2030. It is seen as the desired option for electric power, given its relative efficiency and environmental advantages in comparison with other fossil energy sources. Natural gas burns more cleanly than either coal or oil, making it a more attractive choice for countries seeking to reduce greenhouse gas emissions.

Natural Gas Demand

☛ 2004—104 Quads

☛ 2030—170 Quads

64% increase

Coal remains a vital fuel for the world's electricity markets, and it is expected to continue to dominate energy markets in developing Asia. World coal consumption increased sharply from 2003 to 2004, largely because of a 17-percent increase in China and India. As a result, coal's share of total world energy use climbed from 25 percent in 2003 to 26 percent in 2004. With oil and natural gas

Coal Demand

☞ 2004—115 Quads

☞ 2030—199 Quads

74% increase

prices expected to continue rising, coal is an attractive fuel for nations with access to ample coal resources, and its share of world energy consumption is projected to increase further, to 28 percent in 2030. In particular, the United States, China, and India are well positioned to displace more expensive fuels with coal, and together the three nations account for 86 percent of the expected increase from 2004 to 2030.

In Japan, Western Europe, Eastern Europe, and the former Soviet Union (excluding Russia), coal is expected to be displaced by natural gas, and in France by nuclear power, for electric power generation. In these regions population growth is slow or declining, electricity demand growth is slow, and natural gas and nuclear power are likely to continue providing significant amounts of electricity.

Electricity

In 2004, the world consumed 16,624 billion kWh. The United States consumed 3,804 billion kWh, followed by China and Japan with 2,080 and 974 billion kWh, respectively. Coal accounted for more than 50 percent of the electricity generation in the United States. Nuclear sources accounted for another 16 percent of the electricity generation. Worldwide net electricity consumption is projected to nearly double between 2004 and 2030, from 16,624 billion kWh to 30,364 billion kWh. Strong growth in electricity use is expected in the countries of the developing world, where electricity demand increases by an average of 3.5 percent per year, compared with a projected average increase of 2.4 percent per year worldwide. Robust economic growth in many of the developing nations is expected to boost demand for electricity to run newly purchased home appliances for air conditioning, cooking, space and water heating, and refrigeration.

Electricity Demand

☞ 2004—16.4 Trillion kWh

☞ 2030—30.4 Trillion kWh

67% increase

Electricity generation is expected to increase from several sources (Figure 2.7). Worldwide, electricity generation from nuclear power is projected to increase from 2,619 billion kWh in 2004 to 3,619 billion kWh in 2030. It was anticipated in past years that nuclear generation would decline in the later years of the projections because aging nuclear reactors (especially in developed countries) would be taken out of operation and not replaced. The role of nuclear power in meeting future electricity demand has been reconsidered more recently, given concerns about rising fossil fuel prices, energy security, and greenhouse gas emissions. On the other hand, issues related to plant safety, radioactive waste disposal, and the proliferation of nuclear weapons, which continue to raise public concerns in many countries, may hinder the development of new nuclear

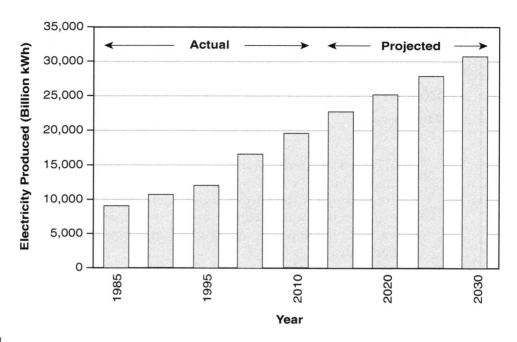

Figure 2.7

Growth in world electricity (projected to 2030).

power reactors. According to International Energy Outlook (2007) projections by the U.S. Department of Energy (U.S. DOE) in Europe only, and particularly in Germany and Belgium, a decline in nuclear power generation after 2010 is projected due to governmental policies. In contrast, developing Asia is poised for a robust expansion of nuclear generation. For example, in China, electricity generation from nuclear power is projected to grow at an average annual rate of 7.7 percent from 2004 to 2030, and in India it is projected to increase by an average of 9.1 percent per year.

Moderate growth in the world's consumption of hydroelectricity and other renewable energy resources is projected over the next 26 years, averaging 1.9 percent per year. Much of the projected growth in renewable generation is expected to result from the completion of large hydroelectric facilities in developing countries, particularly in developing Asia and South and Central America. China, India, and other developing Asian countries are constructing or planning new, large-scale hydroelectric facilities. India has about 12,020 megawatts of hydroelectric capacity under construction. China also has a number of large-scale hydroelectric projects under construction, including the 18,200-megawatt Three Gorges Dam project and the 12,600-megawatt Xiluodu project on the Jisha River.

In Central and South America, Brazil is planning for a number of new hydropower projects that the country hopes to complete to keep up with electricity demand after 2010, including the 3,150-megawatt Santo Antonio and

3,300-megawatt Jirau projects on the Madeira River. Among the industrialized nations, only Canada has plans to construct any sizable hydroelectric projects over the forecast period. Much of the expected increment in renewable energy consumption in the industrialized world is projected to be non-hydropower renewable energy sources, particularly wind energy in Western Europe and the United States. In addition, the use of biomass and geothermal energy sources is expected to grow rapidly in the United States.

United States Energy Consumption

The United States is the world's largest energy producer, consumer, and net importer. It also ranks eleventh worldwide in reserves of oil, sixth in natural gas, and first in coal.

Figure 2.8 shows the total energy flow for the United States for the year 2006; Figure 2.9 shows energy consumption by source. The United States produced 70.95 quads and consumed 97.35 quads, importing 26.4 quads of energy (27.1 percent of the total energy). Fossil fuels accounted for 85.8 percent of the total energy consumed. The four end-users of the energy—residential, commercial, industrial, and transportation—use 21.5 percent, 17.9 percent, 33.3 percent, and 27.2 percent, respectively, of the total energy. Total primary U.S. energy consumption is projected to increase from 98 quadrillion BTUs in 2002

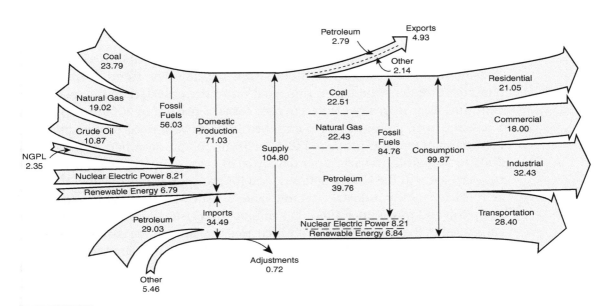

| Figure 2.8 | *Total energy flow (quadrillion BTUs) of the United States (2006).* |

Source: *http://www.eia.doe.gov/emeu/aer/diagram1.jpg*

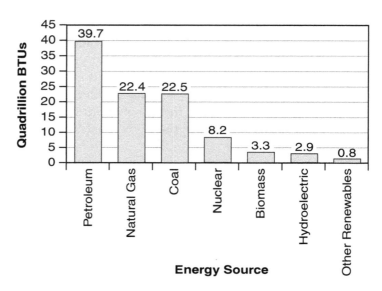

Figure 2.9

United States energy consumption by source (2006).
Data Source: *Energy Information Administration, U.S. DOE, 2007.*

to 137 quadrillion BTUs in 2025. U.S. energy consumption is expected to increase more rapidly than the domestic energy supply, and net imports will constitute 36 percent of consumption in 2025, up from 26 percent in 2002. The complete United States energy foundation is illustrated in Figure 2.10.

Petroleum

In 2005, the domestic production of crude oil was 5.2 million barrels per day, and the imports were 10.1 million barrels per day. The trend of increasing United States dependence on imported oil in the past decade is expected to continue. Net imports, which accounted for 60.5 percent in 2005, are expected to decline to 55 percent by 2015 and then increase back to 61 percent by 2030. Petroleum demand is projected to grow from 20.8 million barrels per day in 2005 to 26.8 million barrels per day in 2030. Approximately 66.5 percent of the petroleum in the United States is used for transportation, and about 23 percent is used by the industrial sector. Table 2.1 shows motor-vehicle fuel consumption and travel since 1960. The number of registered vehicles and the fuel consumed in the United States have increased threefold in the past four decades, but the number of miles traveled by these vehicles has increased fourfold. The average miles traveled per year has increased from 9,500 miles in 1980 to 12,400 miles in 2005. The percentage of people using car pools has decreased from 14 percent to 9.5 percent, as shown in Figure 2.11.

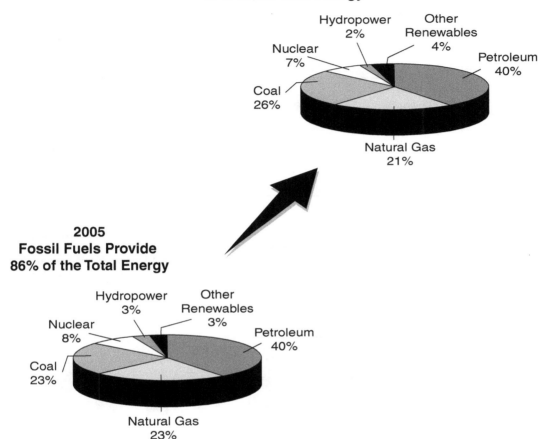

2030
Fossil Fuels Provide
87% of the Total Energy

Hydropower 2%
Other Renewables 4%
Petroleum 40%
Nuclear 7%
Coal 26%
Natural Gas 21%

2005
Fossil Fuels Provide
86% of the Total Energy

Hydropower 3%
Other Renewables 3%
Nuclear 8%
Petroleum 40%
Coal 23%
Natural Gas 23%

Figure 2.10   *Energy foundation of the United States.*

Natural Gas

U.S. natural gas production and consumption were nearly in balance through 1986. After that, consumption began to outpace production, and imports of natural gas rose to meet U.S. requirements for the fuel. In 2006, production stood at 18.5 trillion cubic feet (Tcf), net imports at 3.4 Tcf, and consumption at 21.9 Tcf. Total natural gas consumption in the United States is projected to increase from 22.0 trillion cubic feet in 2005 to 26.1 trillion cubic feet in 2030. Much of the growth is expected before 2020, with demand for natural gas in the electric power sector growing from 5.8 trillion cubic feet in 2005 to a peak of 7.2 trillion cubic

Table 2.1 *Motor-Vehicle Fuel Consumption and Travel*

	1960	1970	1980	1990	2000	2005
Vehicles registered (thousands)[a]	73,858	111,242	161,490	193,057	225,821	241,194
Vehicle-miles traveled (millions)	718,762	1,109,724	1,527,295	2,144,362	2,746,925	2,989,807
Fuel consumed (million gallons)	57,880	92,329	114,960	130,755	[R]162,554	179,100
Average miles traveled per vehicle (thousands)	9.7	10.0	9.5	11.1	12.2	12.4
Average miles traveled per gallon	12.4	12.0	13.3	16.4	16.9	16.7
Average fuel consumed per vehicle (gallons)	784	830	712	677	[R]720	743

KEY: [R] = Revised. [a] = Includes personal passenger vehicles, buses, and trucks.
SOURCE: 1960–94: U.S. Department of Transportation, Federal Highway Administration, *Highway Statistics Summary to 1995*, FHWA-PL-97-009 (Washington, DC: July 1997), table VM-201A. 1995–2005: *Ibid., Highway Statistics* (Washington, DC: Annual issues), table VM-1.

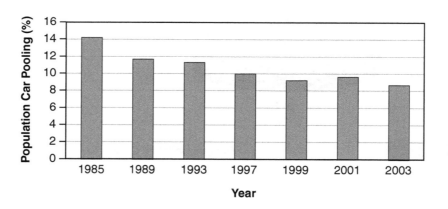

Figure 2.11 *Percentage of population using car pools as a function of year.*

feet in 2020. Natural gas use in the electric power sector will decline after 2020, to 5.9 trillion cubic feet in 2030, as new coal-fired generating capacity displaces natural-gas-fired generation. Continued growth in residential, commercial, and industrial consumption of natural gas will be roughly offset by the projected decline in natural gas demand for electricity generation. As a result, overall natural gas consumption is projected to be almost flat between 2020 and 2030.

With natural gas production in the United States remaining relatively constant, imports of natural gas are projected to rise to meet an increasing share of domestic consumption. Most of the expected growth in U.S. natural gas imports is in the form of liquefied natural gas (LNG). Higher oil prices are expected to reduce world petroleum consumption and increase natural gas consumption. In addition, some LNG contract prices are tied directly to crude oil prices, which could exert upward pressure on LNG prices.

Coal

Coal is another important fossil fuel, accounting for 23 percent of the total primary energy in the United States. Coal consumption in the year 2005 was 1,125 million tons. Almost 92 percent of all coal consumed in the United States was in the electric power sector. According to the Annual Energy Outlook (2007), coal use for electricity generation at existing plants and construction of a few new coal-fired plants are projected to grow at an average rate of 1.1 percent per year from 2005 to 2015. The growth in coal production is projected to be even stronger from 2015 to 2030, averaging 1.8 percent per year, as substantial amounts of new coal-fired generating capacity are added and several Coal to Liquids (CTL) plants are brought on line. The primary reason for the change in the rate of growth is a substantial increase in projected coal demand for electricity generation resulting from higher natural gas prices. Coal is projected to play a more important role in future additions to electricity generation capacity, particularly in the later years of the forecast. It is projected that a total of 1,772 million tons of coal will be consumed in 2030.

Electricity

Total electricity consumption, including purchases from electric power producers and on-site generation, is projected to reach 5,797 billion kilowatt hours in 2025. Growth in electricity use for computers, office equipment, and electrical appliances in the residential and commercial sectors is partially offset by improved efficiency in the effects of demand-side management programs and by slower growth in electricity demand for some applications, such as air conditioning.

In 2005, coal-fired plants accounted for 50 percent of generation and natural-gas-fired plants for 19 percent. Most capacity additions over the next 10 years are natural-gas-fired plants, increasing the natural gas share to 22 percent and lowering the coal share to 49 percent in 2015. As natural gas becomes more expensive, however, more coal-fired plants will be built. In 2030, the generation shares for coal and natural gas are projected to be 57 percent and 16 percent, respectively. Nuclear and renewable generation will increase as new plants are built stimulated by federal tax incentives and rising fossil fuel prices. Nuclear generation is expected to fall from 19 percent in 2005 to 15 percent in 2030. The generation share from renewable capacity (about 9 percent of total electricity supply in 2005) is projected to remain at about 9 percent.

Growth in the Energy Demand

Growth in energy consumption in the world and in the United States as a function of time follows what is known as an exponential function. The exponential increase is characterized as follows: The amount of change (increase in energy consumption) per unit of time is proportional to the quantity (or consumption) at that time.

$$\frac{\Delta N}{\Delta t} \propto N$$

or

$$\frac{\Delta N}{\Delta t} = \lambda N$$

where Greek letter Δ (delta) is the change or increment of the variable and λ (lambda) is the growth rate. After some mathematical methods, it can be shown that the equation changes to the form:

$$N = N_0 e^{\lambda t} \tag{2.1}$$

where N is the projected value at time t, N_0 is the initial value, λ is the rate of increase (as decimal), and t is the projected time. Application of this equation is shown in Illustration 2.1

Illustration 2.1

Total primary energy consumption of the world is 404 quadrillion BTUS in 2002. It is projected that the energy use will increase at the rate of 2.4% annually. What is the projected energy consumption in 2020?

Energy consumption can be predicted using Equation 2.1:

$$N = N_0 e^{\lambda t}$$

Known quantities: $N_0 = 404 \times 10^{15}$ BTUs
$\lambda = 2.4\%$ or 0.024
$t = 18$ years

Therefore,

$$Energy\ Consumption = 404 \times 10^{15}\ e^{(0.024 \times 18)}$$

$$= 622.3 \times 10^{15}\ \text{BTUs or 622 Quads}$$

We can also determine how long it takes for N_0 to become $2N_0$ (twice its original number or double). That time period is called doubling time. After some mathematical steps it can be written as:

$$\text{Doubling time} = \frac{70}{\%\ \text{Growth rate per year}} \qquad (2.2)$$

Illustration 2.2

Use of coal is projected to increase at the rate of 1.7% per year in the United States. How long will it take to double its usage?

$$\text{Doubling time (years)} = \frac{70}{1.7} = 41.17\ \text{years}$$

In 41.17 years, the consumption of coal will be twice as much as it is today.

Energy Reserves

It is evident that the energy requirement is going to increase in the future, making the world and the United States more dependent on fossil fuels. These fossil fuels are non-renewable energy sources with a finite lifetime. So the question is: Will we have enough supply for future energy requirements? The answer to this question depends on the quantity of fossil fuels we have in the ground. Energy sources that have been discovered but not produced cannot be easily measured. They are trapped several feet below the surface and sometimes cannot be measured with precision.

There are several terms used to report the estimates of the energy resources. *Reserves* represent that portion of demonstrated resources that can be recovered economically with the application of extraction technology available currently or in the foreseeable future. Reserves include only recoverable energy. *Resources* represent that portion of the energy that is known to exist or even suspected to exist, regardless of technical or economic viability. Reserves are a subset of resources.

Table 2.2 shows the reserves and consumption of the main fossil fuels in the United States and the world. The United States has 27 percent of the world's coal deposits, but only less than 2 percent and 3 percent of the world's oil and natural gas, respectively. The United States uses about 25 percent of world's coal, oil, and natural gas.

Table 2.2   *United States and World Reserves and Consumption of Fossil Fuels*

	UNITED STATES RESERVES	UNITED STATES ANNUAL CONSUMPTION	WORLD RESERVES	WORLD ANNUAL CONSUMPTION
Petroleum (billions of barrels) as of Jan. 1, 2006	21.8	7.556	1,292.9[a] 1,119.6[b]	30.41
Natural Gas (Dry) Trillion Cubic Feet as of Jan. 1, 2006	204.4	22.24	6,124[a] 6,226.6[b]	103.7
Coal (millions of tons) as of Jan. 1, 2005	267,554	1,107	997,748	6,099

[a] *Oil and Gas Journal;* [b] *World Oil*
SOURCES: *International Energy Annual 2005 and 2006*

How Long Will the Reserves Last?

How long these reserves last depends on the rate at which we consume them. For example, let's assume that we have $100,000 in the bank (reserves) and from it we draw $10,000 every year (consumption). The reserves will last for 10 years ($100,000/$10,000 per year). However, in this case we are assuming that we do not add any money to our deposit and we do not increase our withdrawal. This scenario is generally not true in the case of the life of an energy reserve. We may find new reserves, and our energy consumption or production can also increase. In the case of energy reserves, although we know that we might find new resources, we do not know how much we may find. The consumption, however, can be predicted with some accuracy, based on the past rates.

Lifetime of Current Reserves at Constant Consumption

We can calculate the life of current petroleum reserves by dividing the current reserves by current consumption. At the current rate of consumption, the lifetime of the world's petroleum, natural gas, and coal reserves is 39 years, 60 years, and 164 years, respectively. At the current rate of consumption, the current United States petroleum, natural gas, and coal reserves will last approximately 2.9 years, 9.2 years, and 242 years, respectively. It is important to note that this calculation is based on the assumption that the entire United States petroleum consumption is coming from the United States reserves. Because we import more than one half of the consumption, the petroleum reserves at the current rate will last about 11 years. If the consumption increases in the future, the life will be shorter. However, there is also a chance of adding more reserves with more exploration and discoveries or even changes in technology. The increase in consumption can change, depending on the price of petroleum and other alternative fuels. Therefore, these lifetimes are not carved in stone. It can be debated whether the United States reserves last for 6 years, 10 years, or even 20 years, but there is increasing consensus that we must change our lifestyle. We must conserve, innovate (get more with less), or learn to live without these resources.

Sources

Aubrecht, G. L. (1995). *Energy.* Englewood Cliffs, NJ: Prentice Hall.

Christensen, J. W. (1996). *Global science: Energy resources environment* (4th ed.). Dubuque, IA: Kendall/Hunt.

Energy Information Administration. (2007). *Annual Energy Outlook.* DOE/EIA 0383. Washington, D.C.: United States Department of Energy.

Energy Information Administration. (2006). *Annual Energy Review.* DOE/EIA 0384. Washington, D.C.: United States Department of Energy.

Energy Information Administration. (2006). *International Energy Outlook.* DOE/EIA 0484. Washington, D.C.: United States Department of Energy.

Fay, J. A., and Golomb, D. S. (2002). *Energy and the environment.* New York: Oxford University Press.

Hinrichs, R. A. (1992). *Energy.* Philadelphia: Saunders College Publishers.

questions

1. Why is the energy use per person in the world increasing?

2. The United States, with 5% of the world's population, uses about 25% of the world's energy. Explain how this can be justified.

3. List the reasons why the United States' per-capita energy consumption is higher compared to most other countries in the world.

4. List reasons why the United States' energy consumption per dollar of GDP is higher than most of the industrialized nations.

5. What is the difference between reserves and resources?

6. List the changes that you would make in your personal lifestyle if you were mandated to reduce your energy consumption by 25%.

7. What variables determine the lifetime of a non-renewable resource?

multiple choice questions

1. The United States imports about 30% of all the energy that is consumed in the country.
   a. True
   b. False

2. Per dollar of GDP per year, the United States uses the most energy (BTUs/yr) of any country in the world.
   a. True
   b. False

3. Petroleum reserves for the U.S. at the present rate of consumption and imports will last for approximately
   a. 200 years
   b. 40 years
   c. 100 years
   d. 10 years

4. The U.S. has the highest crude oil reserves in the world.
   a. True
   b. False

5. Reserves represent that portion of demonstrated resources that can be recovered with the application of extraction technology available currently or in the foreseeable future irrespective of economic viability.
   a. True
   b. False

6. Natural gas meets a majority of the U.S. energy demand more than any other fuel.
   a. True
   b. False

7. Natural gas use is expected to decrease in the next two decades because it is one of the fossil fuels that cause damage to the environment.
   a. True
   b. False

8. Renewable energy sources meet ____% of the U.S. energy demand.
   a. 50%
   b. 8%
   c. 25%
   d. 75%

9. In the next 20 years, it is expected that renewable energy sources will supply more than half of the U.S. energy requirements.
   a. True
   b. False

10. Which fuel supplies the greatest proportion of primary energy in the U.S.?
    a. Coal
    b. Petroleum
    c. Natural Gas
    d. Nuclear

11. Coal reserves for the U.S. at the present rate of consumption will last for about
    a. 200 years
    b. 40 years
    c. 100 years
    d. 10 years

12. United States imports (net) about ____% of its petroleum needs.
    a. 5%   c. 60%
    b. 25%   d. 70%

Energy Supply and Demand Puzzle

Across

2. Most of the coal produced in this country is used to generate this form of energy.

5. Most of the energy in this country is used by this sector.

7. Energy growth follows this mathematical function.

8. Most of the petroleum is used by this end sector.

10. Most of the sulfur in the fuels is emitted as this gas.

11. The United States imports over 50% of this energy source.

12. These fuels account for about 85% of the total energy use in the United States.

Down

1. Time required to double the original quantity is called this.

3. Heavy use of fossil fuels in the U.S. makes this country the largest emitter of this gas.

4. Fifty percent of the electricity generated in the U.S. in the year 2004 came from this primary energy source.

6. The majority of this primary energy source produced in the U.S. is used by the industrial sector.

9. In addition to sulfur dioxide, this also contributes to the acid rain problem.

Energy Efficiency

goals

- **To define and calculate the efficiency of an energy conversion device**

- **To understand and articulate the concept of entropy**

- **To understand and explain the operating principles of a heat engine**

- **To calculate overall efficiency from step efficiencies**

$\mathcal{I}$n Chapter 1, we have seen that energy can be transformed from one form to another, and that, during this conversion, all the energy that we put into a device comes out. However, all the energy that we put in may not come out in the desired form. For example, we put electrical energy into a bulb and the bulb produces light (which is the desired form of output from a bulb), but we also get heat from the bulb (undesired form of energy from an electric bulb). Figure 3.1 summarizes energy flow into and out of any energy conversion device.

When all forms of energy coming out of an energy conversion device are added up, the sum will be equal to the energy put into a device. Energy output must be equal to the input. This means that energy cannot be destroyed or created—it can only change its form. In the case of an electric bulb, the electrical energy is converted to light and heat. Light is a useful form, but heat is not desired from an electric bulb. This also means that all the energy put in will come out but not all will be in the useful form. So the efficiency can be defined as:

$$Efficiency = \frac{Useful\ energy\ output}{Total\ energy\ input} \qquad (3.1)$$

Illustration 3.1

An electric motor consumes 100 watts [a joule per second (J/s)] of power to obtain 90 watts of mechanical power. Determine its efficiency.

Solution: Input to the electric motor is in the form of electrical energy and the output is mechanical energy.

Using equation 3.1,

$$Motor\ efficiency = \frac{Mechanical\ power}{Electrical\ power} = \frac{90\ \cancel{W}}{100\ \cancel{W}} = 0.9$$

or efficiency is 90%.

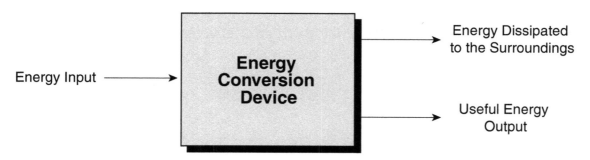

Figure 3.1 *Energy flow diagram for an energy conversion device.*

Illustration 3.1 is a very simple case, because both mechanical and electrical powers are given in watts. Units of both the input and the output must match.

Illustration 3.2

The United States power plants consumed 39.5 quadrillion BTUs of energy and produced 3.675 trillion kWh of electricity. What is the average efficiency of the power plants in the U.S.?

$$Efficiency = \frac{Useful\ energy\ output}{Total\ energy\ output}$$

Total energy input = 39.5×10^{15} BTUs and the Useful energy output is 3.675×10^{12} kWh. Recall that both units must be the same. So we need to convert kWh into BTUs. Given that 1 kWh = 3412 BTUs,

$$Efficiency = \frac{3.675 \times 10^{12}\ kWh}{39.5 \times 10^{15}\ BTUs} \times \frac{3412\ BTUs}{1\ kWh} = 0.3174\ or\ 31.74\%$$

These efficiencies are not 100 percent, and sometimes they are fairly low. Table 3.1 shows typical efficiencies of some of the devices used in day-to-day life.

From our discussion of national and global energy usage patterns in the previous chapter, we have seen that about 40 percent of U.S. energy is used in power generation, and the efficiency of a power plant is about 35 percent. Transportation accounts for another 27 percent of the energy use in the United States.

Table 3.1   *Typical Efficiencies of Some of the Commonly Used Devices*

DEVICE	EFFICIENCY
Electric motor	90%
Home gas furnace	95%
Home oil furnace	80%
Home coal stove	75%
Steam boiler in a power plant	90%
Overall power plant	35%
Automobile engine	25%
Electric bulb	
Incandescent	5%
Fluorescent	20%

The efficiency of automobiles is about 25 percent. Thus, over 67 percent of the total primary energy in the United States is used in relatively inefficient conversion processes. Why are power plant and automobile design engineers allowing this? Can they do better? There are some natural limitations when converting energy from heat to work.

What Is Thermal Energy and How Is It Measured?

Thermal energy is energy associated with the random motion of molecules. It is indicated by temperature, which measures the relative warmth or coolness of an object. A temperature scale is determined by choosing two reference temperatures and dividing the temperature difference between these two points into a certain number of degrees.

The two reference temperatures used for most common scales are the melting point of ice and the boiling point of water. On the Celsius (or Centigrade) temperature scale, the melting point is taken as 0°C and the boiling point as 100°C, and the difference between them is divided into 100°C. On the Fahrenheit temperature scale, the melting point is taken as 32°F and the boiling point as 212°F, with the difference between them equal to 180°F. The temperature of a substance does not measure its heat content but rather the average kinetic energy of its molecules resulting from their motions. A 6-ounce cup of hot water and a 12-ounce cup of hot water at the same temperature do not have the same heat content. Because they are at the same temperature, the average kinetic energy of the molecules is the same; however, the water in the 12-ounce cup has more molecules than that in the 6-ounce cup and thus has greater heat energy.

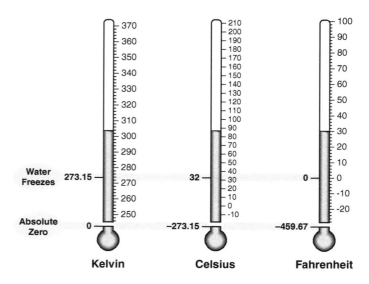

Temperature scales.

Figure 3.2

When water molecules freeze at 0°C, the molecules still have some energy compared to ice at −50°C. What is the temperature at which all the molecules absolutely have zero energy? A temperature scale can be defined theoretically for which 0° corresponds to zero average kinetic energy. Such a point is called absolute zero, and such a scale is known as an absolute temperature scale. At absolute zero, the molecules do not have any energy. The Kelvin temperature scale is an absolute scale having degrees the same size as those of the Celsius temperature scale. Figure 3.2 shows all three temperature scales on one scale.

Therefore, all the temperature measurements related to energy measurements must be made on the Kelvin scale. Let's look at those automobile energy conversions (Figure 3.3).

Any device that converts thermal energy into mechanical energy is called a heat engine. In these devices, such as automobiles or power plants, high temperature heat (thermal energy) produced by burning a fuel is partly converted to mechanical energy, and the rest is rejected into the atmosphere, as shown in Figure 3.4.

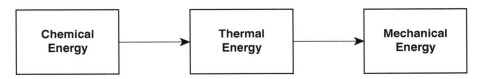

Figure 3.3   *Energy conversions in an automobile.*

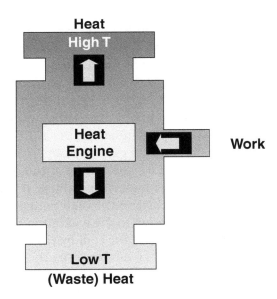

Figure 3.4   *Energy flow in a heat engine.*

A general expression for the efficiency of a heat engine can be written as:

$$Efficiency = \frac{Work}{Heat\ energy_{Hot}}$$

We know that all the energy that is put into the engine has to come out either as work or waste heat. So work is equal to Heat at High temperature minus Heat rejected at Low temperature. Therefore, this expression becomes:

$$Efficiency = \frac{Q_{Hot} - Q_{Cold}}{Q_{Hot}} \tag{3.2}$$

where Q_{Hot} = Heat input at high temperature and Q_{Cold} = Heat rejected at low temperature. With the symbol η (Greek letter eta), used for efficiency, this expression can be rewritten as:

$$\eta(\%) = \left(1 - \frac{Q_{Cold}}{Q_{Hot}}\right) \times 100 \tag{3.3}$$

Equation 3.3 is multiplied by 100 to express the efficiency as percent. French Engineer Sadi Carnot showed that the ratio of Q_{HighT} to Q_{LowT} must be the same as the ratio of the temperatures of high temperature heat and the rejected low temperature heat. So this equation can be simplified as:

$$\eta = \left(1 - \frac{T_{Cold}}{T_{Hot}}\right) \times 100\% \tag{3.4}$$

This efficiency is the theoretical maximum efficiency one can get when the heat engine is operating between those temperatures. This efficiency, also called "Carnot Efficiency," depends on the two temperatures—the temperature at which the high and low temperature reservoirs operate. In the case of an automobile, they are the temperature of the combustion gases inside the engine (T_{Hot}) and the temperature at which the gases are exhausted from the engine (T_{Cold}). When the exhaust leaves at a higher temperature, it carries more energy out, and that energy is not available to be converted to work. Therefore, the higher the T_{Cold}, the lower the efficiency. Similarly, if the T_{Hot} is increased by increasing the temperature of the combustion gases, we can get higher efficiencies.

Why, then, should we operate the automobiles at low efficiencies? It is not that we cannot achieve high temperatures; it is that we do not have the engine materials that can withstand the high temperatures. As a matter of fact, we do not let the engine gases go to the maximum temperature that they can go even now; we try to keep the engine cool by circulating the coolant. So we are taking the heat out of the gases (thus lowering the T_{Hot}) and making the engine operate at cooler temperatures so that the engine is protected, but in doing so we are lowering the efficiency of an automobile.

From Illustration 3.3, it can be inferred that a maximum of 64 percent of the fuel energy can go to generation. To make the Carnot efficiency as high as possible, either T_{Hot} should be increased or T_{Cold} (temperature of heat rejection) should be decreased.

Illustration 3.3

For a coal-fired utility boiler, the temperature of high-pressure steam would be about 540°C and T_{Cold}, the cooling tower water temperature, would be about 20°C. Calculate the Carnot efficiency of the power plant.

Solution: Carnot efficiency depends on high temperatures and low temperatures between which the heat engine operates. We are given both temperatures, however, the temperatures need to be converted to Kelvin:

$$T_{Hot} = 540°C + 273 = 813 \text{ K}$$

$$T_{Cold} = 20°C + 273 = 293 \text{ K}$$

$$Carnot\ efficiency = \left(1 - \frac{293\,K}{813\,K}\right) \times 100 = 64\%$$

Entropy and Quality of Energy

Entropy is in a way "disorder." In a high temperature system, molecules or atoms have more energy, and they move randomly at high velocities, i.e. disorder. Steam at high temperature, therefore, has high entropy. On the other hand, the motion of a turbine or piston in an engine (mechanical energy) is low entropy or a low disorder system because of the sense of direction or organized motion. In order to convert a highly disordered system to an ordered system one needs to expend (waste) energy. It is analogous to cleaning up a dirty room. We need to spend energy and effort to convert a high entropy room into a low entropy room. A heat engine converts heat energy (high entropy system) into mechanical energy (low entropy system), and therefore a lot of energy is spent (wasted) in the process, thus reducing the efficiency. It can also be inferred that thermal energy is low quality energy and mechanical energy is high quality.

Energy Conversions in a Power Plant

Coal, oil, and gas have chemical energy stored in the chemical bonds of the fuel. Figure 3.5 illustrates a power plant. When the fuel is burned, the chemical bonds

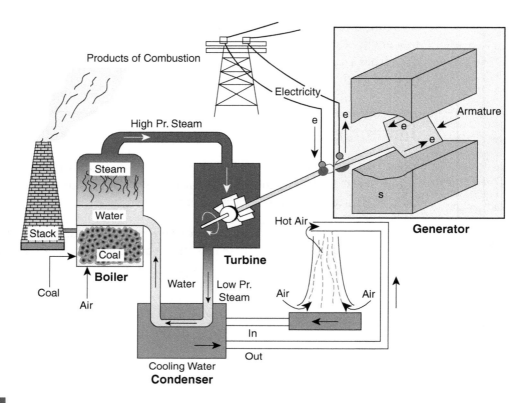

Products of Combustion

Electricity

Armature

High Pr. Steam

Steam

Water

Stack

Coal

Boiler

Coal

Air

Water

Turbine

Low Pr.
Steam

Hot Air

Generator

Air

Air

In

Out

Cooling Water
Condenser

Figure 3.5

Schematic of a power plant.

are broken and new bonds are formed, releasing thermal energy. This thermal energy is transferred to water that turns into high-pressure, high-temperature steam, which turns the turbine and converts the thermal energy into mechanical energy. The steam, after turning the turbine, will still have some energy, but not enough to turn the turbine. The low-pressure steam is condensed into water, and the water is sent back to the boiler. The turbine is connected to a generator, and in the generator the mechanical energy is converted into electricity by turning a conductor in a magnetic field.

The basic energy conversions in the three main components in a power plant are shown in Figure 3.6 quantitatively.

Let's say the boiler takes in 100 BTUs of chemical energy and produces 88 BTUs of useful thermal energy. The 88 BTUs of thermal energy from the boiler go into the turbine, and 36 BTUs equivalent of mechanical energy (movement of turbine blades) are produced. These 36 BTUs of mechanical energy are transferred to the generator, which converts them to 10.26 Wh of electrical energy. Using the energy efficiency concept, we can calculate the component and overall efficiency:

$$Overall\ efficiency = \frac{Electrical\ energy\ output}{Chemical\ energy\ input}$$

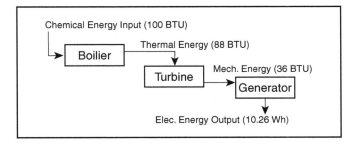

Figure 3.6

Energy conversions in a power plant.

Here the electrical energy is given in Wh and the chemical energy in BTUs. So Wh can be converted to BTUs, knowing that there are 3.412 Wh in a BTU:

$$Overall\,efficiency = \frac{10.26\,\cancel{Wh}}{100\,\cancel{BTUs}} \times \underbrace{\frac{3.412\,\cancel{BTUs}}{1\,\cancel{Wh}}}_{Conversion\,of\,Wh\,to\,BTUs} = \frac{35\,BTUs}{100\,BTUs} \times 100 = 35\%$$

This overall efficiency can also be expressed in steps as follows:

$$Overall\,efficiency = \underbrace{\left(\frac{Thermal\,energy}{Chemical\,energy}\right)}_{Efficiency\,of\,the\,Boiler} \times \underbrace{\left(\frac{Mechanical\,energy}{Thermal\,energy}\right)}_{Efficiency\,of\,the\,Turbine} \times \underbrace{\left(\frac{Electrical\,energy}{Mechanical\,energy}\right)}_{Efficiency\,of\,the\,Generator}$$

$$Overall\,efficiency = Boiler\,\eta \times Turbine\,\eta \times Generator\,\eta$$

Applying this method to the above power plant example:

$$Overall\,efficiency = \frac{88\,BTUs}{100\,BTUs} \times \frac{36\,BTUs}{88\,BTUs} \times \frac{35\,BTUs}{36\,BTUs}$$

$$= 0.88 \times 0.41 \times 0.97$$

$$= 0.35\,or\,35\%$$

This calculation illustrates that the overall efficiency of a system is equal to the product of the efficiencies of the individual subsystems or processes. What is the implication? We have been looking at the efficiencies of an automobile or a power plant individually. But when the entire chain of energy transformations—from the moment the coal is brought to the surface to the moment the electricity turns into its final form—true overall efficiency of the energy utilization will be revealed (Figure 3.7). The final form at home could be light from a bulb or sound from a stereo.

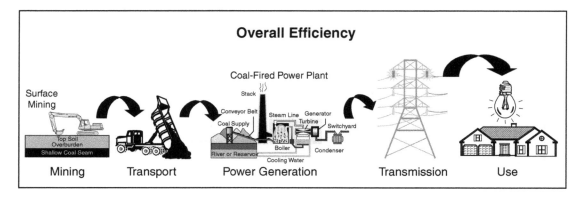

Figure 3.7   *Sequence of steps in converting chemical energy in the ground to light in a room.*

The series of steps shown in Figure 3.7 is:

1. Production of coal
2. Transportation to power plant
3. Electricity generation
4. Transmission of electricity
5. Conversion of electricity into light

If the efficiency of each step is known, we can calculate the overall efficiency of the production of light from coal in the ground. Table 3.2 illustrates the calculation of the overall efficiency of a light bulb.

To generate 6.2 units of light from a relatively efficient fluorescent bulb, we clearly used up 100 units of energy from coal from the ground. This also means that, during various conversion steps, 93.8 units of energy are dissipated into the environment.

Table 3.2   *Overall Efficiency of a Light Bulb*

STEP	STEP EFFICIENCY	CUMULATIVE EFFICIENCY OR OVERALL EFFICIENCY
Extraction of coal	96%	96%
Transportation	98%	94% (0.96 × 0.98) * 100
Electricity generation	35%	33% (0.96 × 0.98 × 0.38 * 100)
Transmission of electricity	95%	31%

Lighting
    Incandescent bulb
    Fluorescent
	
	5%
	20%
		
		1.5%
		6.2%

A similar analysis on automobiles, shown in Figure 3.8 and Table 3.3, indicates that only about 10 percent of the energy in the crude oil in the ground is in fact turned into mechanical energy that moves people.

Sources

Hinrichs, R. A. (1992). *Energy.* Philadelphia: Saunders College Publishers.

Aubrecht, G. L. (1995). *Energy.* Englewood Cliffs, NJ: Prentice Hall.

Fay, J. A., and Golomb, D. S. (2002). *Energy and the environment.* New York: Oxford University Press.

Christensen, J. W. (1996). *Global science: Energy resources environment* (4th ed.). Dubuque, IA: Kendall/Hunt.

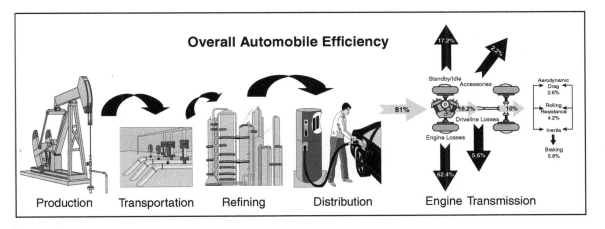

Figure 3.8   *Sequence of steps in converting chemical energy in crude oil in the ground to movement of a car.*

Table 3.3   *Overall Efficiency of a Light Bulb*

STEP	STEP EFFICIENCY	OVERALL OR CUMULATIVE EFFICIENCY
Production of crude	96%	96%
Refining	87%	84%
Transportation	97%	81%
Engine	25%	20%
Transmission	50%	10%

questions

1. A heat engine has a Carnot efficiency of 30%. Useful output from the engine is 1000 J. How much heat is wasted (in Joules)?

2. How can we improve the Carnot efficiency of a heat engine by changing the hot and cold reservoir temperatures?

3. Most of the energy conversion devices that we use in our day-to-day life can be classified as heat engines. Give two examples.

4. The following diagram shows the energy flow to and from a furnace.

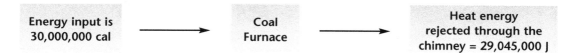

Calculate the efficiency of the furnace.

multiple choice questions

1. Approximately what percentage of electricity does an incandescent light bulb convert into visible light?
   a. 5
   b. 20
   c. 40
   d. 90

2. The flame temperature in an automobile is 1,011°C, and the exhaust is emitted at 74°C. What is the Carnot efficiency?
   a. 7.32%
   b. 72.98%
   c. 92.68%
   d. 27.02%

3. The following step efficiencies apply to the use of gasoline in a car: Crude production: 96%, Refining 87%, Transportation 97%, Engine efficiency 25%. What is the total efficiency of the process?
   a. 40%
   b. 30%
   c. 20%
   d. 50%

4. The flame temperature in an automobile is 1,000°C, and the exhaust is emitted at 70°C. What is the Carnot efficiency?
   a. 25%
   b. 65%
   c. 73%
   d. 33%

5. If the energy input of a system is 50 calories and the output is 25 calories, what is the system efficiency?
   a. 100%
   b. 50%
   c. 200%
   d. 25%

6. Heat engines are inefficient because the energy conversion is
   a. from low entropy to high entropy
   b. from high entropy to low entropy
   c. from low temperature to high temperature
   d. from high temperature to low temperature

7. Energy can be created when we burn fossil fuels.
   a. True
   b. False

8. Assuming that an automobile engine's operating temperature remains the same, its Carnot efficiency increases in
   a. Summertime
   b. Wintertime

9. The useful output from a heat engine is 238 cal. The energy that is wasted is 5,667 J. What is the Carnot efficiency of the engine?
   a. 4%
   b. 15%
   c. 17.6%
   d. none of the above

10. Three energy conversion processes take place in succession. The first has an efficiency of 50%, the second is 40% efficient, and the third 5%. What is the overall efficiency (%) of the entire process?
    a. 10
    b. 40
    c. 50
    d. 1

11. The turbine is the _____ efficient component in a power plant.
    a. least
    b. most

12. The flame temperature in an automobile is 1,650°F, and the exhaust is emitted at 212°F. What is the Carnot efficiency?
    a. 88.8%
    b. 87.2%
    c. 68.2%
    d. 25%

13. The function of a generator in a power plant is to convert
    a. chemical energy to mechanical energy
    b. mechanical energy to electrical energy
    c. thermal energy to mechanical energy
    d. none of the above

14. Which of these is false?

    a. $Efficiency = \dfrac{UsefulE}{TotalE}$

    b. $Efficiency = \dfrac{Total - WastedE}{WastedE + UsefulE}$

    c. $Efficiency = \dfrac{UsefulE}{WastedE + UsefulE}$

    d. $Efficiency = 1 - \dfrac{UsefulE}{WastedE + UsefulE}$

15. The following step efficiencies apply to the use of gasoline in a car: Crude production 92%, Refining 83%, Transportation 97%, Engine efficiency 23.8%. What is the total efficiency of the process?
    a. 73.95%
    b. 17628469.6%
    c. 2.96%
    d. 17.63%

16. In a coal-fired power plant, _____ energy in the fuel is finally converted into _____ energy.
    a. chemical, mechanical
    b. chemical, electrical
    c. mechanical, electrical
    d. mechanical, thermal

17. Which of the following devices is least energy-efficient?
    a. power plant
    b. electric motor
    c. light bulb

18. Total energy output must be equal to the input.
    a. True
    b. False

19. Most energy-conversion processes produce _____ as a byproduct.
    a. light
    b. heat
    c. sound
    d. motion

Energy and the Environment

chapter 4

goals

- ☞ *To gain familiarity with fossil fuel composition*

- ☞ *To understand basic combustion chemistry*

- ☞ *To know the quantitative implications of fossil fuel combustion*

- ☞ *To appreciate the health and environmental effects of products of combustion*

- ☞ *To gain basic understanding of the effects of primary and secondary pollutants*

The earlier discussion on the world and the U.S. energy supply clearly established that the dependence on fossil fuels is high (about 86 percent of the total energy), and this dependence is likely to increase in the next two decades. In this section, we are going to look at what the fossil fuels are and the consequences when these fossil fuels are burnt. These fuels—coal, oil, natural gas—that we primarily depend on were formed over millions of years by the compression of organic material (plant and animal sources), prevented from decaying, and buried in the ground. Fossil fuels are hydrocarbons comprised primarily of carbon and hydrogen and some sulfur, nitrogen, oxygen, and mineral matter. Compositional differences between the varieties of coal, petroleum, and natural gas are shown in Table 4.1.

Mineral matter turns into ash when burnt. The composition and the amounts of these elements change for different fossil fuels (coal, petroleum, and natural gas), but the elements are the same. For example, there is more hydrogen in liquid fuels than in coal per unit mass. Combustion is the rapid oxidation of these constituents that generates heat. When these elements oxidize (or combine with oxygen), products of combustion are formed as shown in Figure 4.1. Although the figure shows coal as the fuel, the discussion applies to any fossil fuel since the elements are the same in all the fossil fuels.

Products of Combustion

The carbon in the fuel combines with the oxygen in the air used for burning and forms carbon dioxide. If there is not enough oxygen for complete oxidation, carbon monoxide (CO) may form. Hydrogen (H) in the fuel oxidizes by combining with oxygen (O) and forms water (H_2O).

Similarly sulfur turns into sulfur dioxide (SO_2) and nitrogen into nitric oxide (NO) and nitrogen dioxide (NO_2). The inorganic minerals turn into ash particles. Some of the fuel (hydrocarbon) may not completely burn and therefore is

Table 4.1   *Composition of Fossil Fuels (wt. %)*

ELEMENT	COAL	PETROLEUM	NATURAL GAS
Carbon	60–96%	85–90%	75%
Hydrogen	2–6%	9–15%	25%
Nitrogen	1–2%	0–0.1%	Traces
Sulfur	0.5–5%	Ppm–4%	Traces
Oxygen	1–30%	Ppm	—

Ppm = parts per million

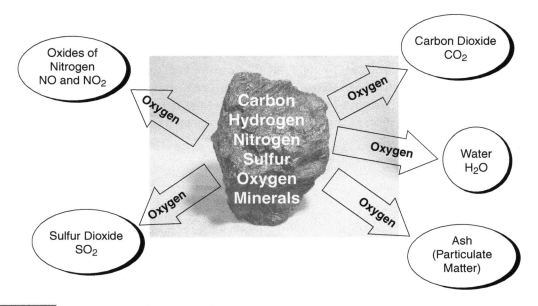

Figure 4.1

Schematic of fossil fuel combustion and product formation.
Courtesy of Execustaff Composition Services.

released into the atmosphere along with the products. The products that are formed during combustion of fossil fuels are shown in Figure 4.2.

Carbon Dioxide (CO_2)

Carbon dioxide is the principal product of the combustion of fossil fuels since carbon accounts for 60–90 percent of the mass of fuels that we burn.

The United States continues to be the largest single emitter of fossil fuel-related CO_2 emissions, reaching an all-time high of 5,912 million metric tons of

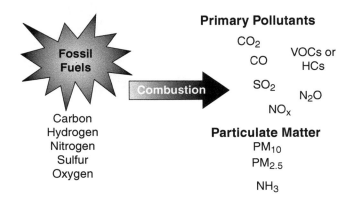

Figure 4.2

Pollutant formation during fossil fuel combustion.

chapter **4** *Energy and the Environment* **57**

carbon dioxide in 2004. Each of the end sectors (users) and their contribution (percent) to the overall CO_2 emissions is shown in Figure 4.3. Per capita values in excess of 19 metric tons of carbon dioxide per person are the highest of the industrialized world. In 2004, 43.1 percent of U.S. CO_2 emissions came from the consumption of petroleum products, and coal usage accounted for 36 percent. About 20.8 percent of the CO_2 emissions were natural gas use. Figure 4.4 shows the contribution of the CO_2 emissions by selected countries.

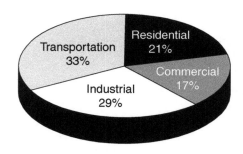

Figure 4.3   *Sources of CO_2 in the United States.*

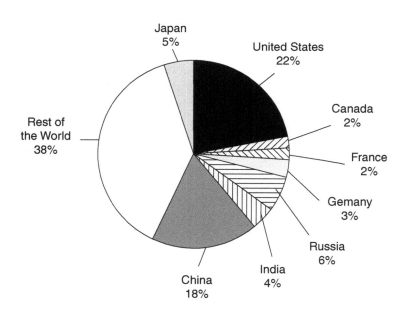

Figure 4.4   *Global CO_2 emissions = 27,044 million tons.*

Source: *Energy Information Administration, "International Energy Annual 2004" (May–July 2006), Table H.1co2*

Carbon Monoxide (CO)

Carbon monoxide, or CO, is a colorless, odorless gas that is formed when carbon in fuel is not burned completely. It is estimated that the United States emitted 112 million tons of CO in the year 2002. Figure 4.5 shows the contribution of various sources to the emissions of CO. It is a component of motor vehicle exhaust, which contributes about 55 percent of all CO emissions nationwide. Other non-road engines and vehicles (such as construction equipment and boats) contribute about 22 percent of all CO emissions nationwide. Higher levels of CO generally occur in areas with heavy traffic congestion. In cities, 85 to 95 percent of all CO emissions may come from motor vehicle exhaust. Other sources of CO emissions include industrial processes (such as metals processing and chemical manufacturing), residential wood burning, and natural sources such as forest fires. Woodstoves, gas stoves, cigarette smoke, and unvented gas and kerosene space heaters are sources of CO indoors. The highest levels of CO in the outside air typically occur during the colder months of the year when inversion conditions are more frequent. An inversion is an atmospheric condition that occurs when the air pollutants are trapped near the ground beneath a layer of warm air.

Sulfur Dioxide (SO$_2$)

Sulfur dioxide, or SO$_2$, belongs to the family of sulfur oxide gases (SO$_x$). Sulfur is prevalent in all raw materials, including crude oil, coal, and ores that contain

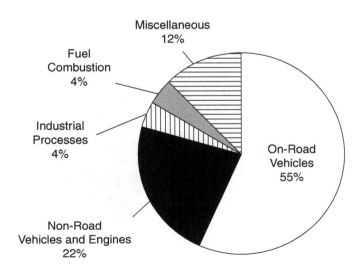

Figure 4.5

Sources of carbon monoxide emissions.

common metals, such as aluminum, copper, zinc, lead, and iron. SO_x gases are formed when fuel containing sulfur, such as coal and oil, is burned, and when gasoline is extracted from oil or metals are extracted from ore. SO_2 dissolves in water vapor to form acid and interacts with other gases and particles in the air to form sulfates and other products that can be harmful to people and their environment. It is estimated that 15.4 million tons of SO_2 are emitted in the United States in the year 2002.

Figure 4.6 shows that over 70 percent of SO_2 released into the air, or more than 10 million tons per year, come from electric utilities, especially those burning coal. Other sources of SO_2 are industrial facilities that derive their products from raw materials like metallic ore, coal, and crude oil, or that burn coal or oil to produce heat for industrial applications. Examples are petroleum refineries, cement manufacturing, and metal processing facilities. Also, locomotives, large ships, and some non-road diesel equipment currently burn high sulfur fuel and emit large quantities of SO_2 into the air.

Nitrogen Oxides (NO_x)

Nitrogen oxides is the generic term for a group of highly reactive gases, all of which contain nitrogen and oxygen in varying amounts. Nitric oxide (NO) and nitrogen dioxide (NO_2) are together called NO_x. Many of the nitrogen oxides are colorless and odorless. However, one common pollutant, nitrogen dioxide (NO_2), can often be seen along with particles in the air as a reddish-brown layer over many urban areas.

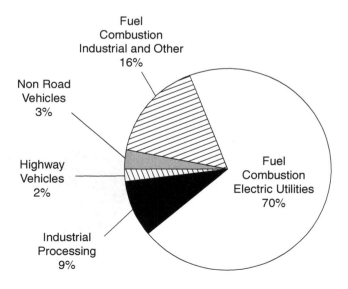

| Figure 4.6 |

Sources of sulfur dioxide emissions.

Nitrogen oxides form when fuel is burned at high temperatures, as in a combustion process. In the United States about 21.1 million tons of NO_x were emitted in 2002. The primary sources of NO_x are motor vehicles, electric utilities, and other industrial, commercial, and residential sources that burn fuels (Figure 4.7).

Lead (Pb)

Lead is a metal found naturally in the environment as well as in manufactured products. The major sources of lead emissions have historically been motor vehicles (such as cars and trucks) and industrial sources. Due to the phase out of leaded gasoline, metals processing is the major source of lead emissions in the air today. The highest levels of lead in air are generally found near lead smelters (devices that process lead ores). Other stationary sources are waste incinerators, utilities, and lead-acid battery manufacturers.

Particulate Matter (PM)

Particulate matter (PM) is the general term used to describe a mixture of solid particles and liquid droplets found in the air. Some particles are large enough to be seen as dust or dirt. Others are so small they can be detected only with an electron microscope. $PM_{2.5}$ describes the "fine" particles that are less than or equal to 2.5 µm (micro meter) in diameter. "Coarse fraction" particles are greater than 2.5 µm, but less than or equal to 10 µm in diameter. PM_{10} refers to all particles less than or equal to 10 µm in diameter (about one-seventh the diameter of a human hair). PM can be emitted directly or formed in the atmosphere.

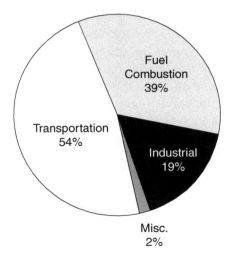

Sources of nitrogen oxides.

Figure 4.7

chapter 4   *Energy and the Environment*   61

"Primary" particles, such as dust from roads or black carbon (soot) from combustion sources, are emitted directly into the atmosphere. "Secondary" particles are formed in the atmosphere from primary gaseous emissions. Examples of secondary particles are sulfates formed from SO_2 emissions from power plants and industrial facilities; nitrates formed from NO_x emissions from power plants, automobiles, and other combustion sources; and carbon formed from organic gas emissions from automobiles and industrial facilities. The chemical composition of PM depends on location, time of year, and weather. Generally, coarse PM is composed largely of primary particles, and fine PM contains more secondary particles.

Health and Environmental Effects of the Primary Pollutants

Carbon Dioxide (CO_2)

Carbon dioxide (CO_2) is not a pollutant that would harm our health, but it is a proven greenhouse gas. It has an ability to absorb infrared radiation that is escaping from the surface of the earth and cause the atmosphere to warm up. Excessive emission of CO_2 along with other greenhouse gases is thought to contribute to the undesirable climate change.

Carbon Monoxide (CO)

Carbon monoxide is a colorless, odorless, and tasteless gas. Hemoglobin (iron compound) in the blood carries the oxygen from the lungs to various tissues and transports CO_2 back to the lungs. Hemoglobin has 240 times more affinity toward CO. Therefore, the hemoglobin reacts with CO and makes that much hemoglobin unavailable for the transport of O_2. This reduces oxygen delivery to the body's organs and tissues. The health threat from levels of CO sometimes found in the ambient air is most serious for those who suffer from cardiovascular disease such as angina pectoris. At much higher levels of exposure not commonly found in ambient air, CO can be poisonous, and even healthy individuals can be affected. Visual impairment, reduced work capacity, reduced manual dexterity, poor learning ability, and difficulty in performing complex tasks are all associated with exposure to elevated CO levels.

Sulfur Dioxide (SO_2)

High concentrations of SO_2 can result in temporary breathing impairment for asthmatic children and adults who are active outdoors. Short-term exposures of asthmatic individuals to elevated SO_2 levels during moderate activity may result in breathing difficulties that can be accompanied by symptoms such as wheezing, chest tightness, or shortness of breath. Other effects that have been associated with longer-term exposures to high concentrations of SO_2, in conjunction with high levels of PM, include aggravation of existing cardiovascular disease, respiratory illness, and alterations in the lungs' defenses. The subgroups of the population that may be affected under these conditions include individuals with heart or lung disease, as well as the elderly and children.

Together, SO_2 and NO_x are the major precursors to acidic deposition (acid rain), which is associated with the acidification of soils, lakes, and streams and the accelerated corrosion of buildings and monuments. We will talk more about this in the next section. SO_2 also is a major precursor to $PM_{2.5}$, which is a significant health concern and a main contributor to poor visibility.

Nitrogen Oxides (NO_x)

Nitric oxide (NO) and nitrogen dioxide (NO_2) together are represented by NO_x. Most of the emissions from combustion devices (approximately 90 percent) are in the form of NO. Short-term exposures (e.g., less than three hours) to low levels of NO_2 may lead to changes in airway responsiveness and lung function in individuals with preexisting respiratory illnesses. These exposures may also increase respiratory illnesses in children. Long-term exposures to NO_2 may lead to increased susceptibility to respiratory infection and may cause irreversible alterations in lung structure. NO_x reacts in the air to form ground-level ozone and fine particulates, which are associated with adverse health effects.

NO_x contributes to a wide range of environmental effects directly, and when combined with other precursors it can result in acid rain and ozone. Increased nitrogen inputs to terrestrial and wetland systems can lead to changes in plant species composition and diversity. Similarly, direct nitrogen inputs to aquatic ecosystems such as those found in estuarine and coastal waters (e.g., Chesapeake Bay) can lead to eutrophication (a condition that promotes excessive algae growth, which can lead to a severe depletion of dissolved oxygen and increased levels of toxins harmful to aquatic life). Nitrogen, alone or in acid rain, also can acidify soils and surface waters. Acidification of soils causes the loss of essential plant nutrients and increased levels of soluble aluminum that are toxic to plants. Acidification of surface waters creates conditions of low pH and levels of aluminum that are toxic to fish and other aquatic organisms. NO_x also contributes to visibility impairment.

Particulate Matter (PM)

Particles smaller than or equal to 10 μm (micro meter or millionth of a meter) in diameter can get into the lungs and can cause numerous health problems.

They have been linked with illness and death from heart and lung disease. Various health problems have been associated with long-term (e.g., multi-year) exposures as well as daily and, potentially, peak (e.g., one-hour) exposures to particles. Particles can aggravate respiratory conditions, such as asthma and bronchitis, and have been associated with cardiac arrhythmias (heartbeat irregularities) and heart attacks. Particles of concern can include both fine and coarse-fraction particles, although fine particles have been more clearly linked to the most serious health effects. Particles larger than 2 micrometers (μm) do not penetrate beyond the nasal cavity or trachea. Particles smaller than 0.1 μm tend to deposit in the tracheobronchial tree and are removed when exhaling. Particles between 0.1 and 2.0 μm penetrate deep into the lungs and settle in respiratory bronchioles or alveolar sacs. People with heart or lung disease, the elderly, and children are at highest risk from exposure to particles.

In addition to health problems, PM is the major cause of reduced visibility in many parts of the United States by scattering and absorbing some of the light emitted or reflected by the body and reducing the contrast. Airborne particles can also impact vegetation and ecosystems and can cause damage to paints and building materials.

Lead

Exposure to lead occurs mainly through inhalation of air and ingestion of lead in food, water, soil, or dust. It accumulates in the blood, bones, and soft tissues and can adversely affect the kidneys, liver, nervous system, and other organs. Excessive exposure to lead may cause neurological impairments such as seizures, mental retardation, and behavioral disorders. Even at low doses, lead exposure is associated with damage to the nervous systems of fetuses and young children, resulting in learning deficits and lowered IQ. Recent studies indicated that lead may be a factor in high blood pressure and subsequent heart disease. Lead can also be deposited on the leaves of plants, presenting a hazard to grazing animals and humans through ingestion.

Secondary Pollutants

The pollutants that are emitted directly from a combustion process are called "primary pollutants." These primary pollutants when emitted into the atmosphere combine with other reactants and form "secondary" pollutants. An example of this would be ozone. When hydrocarbons are emitted and they react with NO_x in the presence of sunlight, they form ozone. Hence, it is a secondary pollutant. Health and environmental effects of secondary pollutants are discussed in the section on the global and regional effects of pollutants.

Global and Regional Effects of Secondary Pollutants

The Earth is continuously receiving energy from the sun. Energy also leaves the Earth in the nighttime (of course in the form of invisible infrared energy!), otherwise the Earth would be continuously warming up. This delicate balance between the energy coming in and leaving due to the natural *greenhouse effect* is what keeps the planet warm enough for us to live on. It is very obvious that if more energy comes into the planet than leaves, it will become warm. Similarly, if the energy that leaves is more than the energy that comes in, the planet will become cool. The atmospheric temperature fluctuates over centuries due to certain natural causes.

Let's try to understand the basic greenhouse effect that makes the Earth habitable. As illustrated in Figure 4.8, energy from the sun reaches the Earth's atmosphere. A part of the energy is reflected by the Earth's cloud cover, and part of it heats up the surface and the atmosphere. In addition, some of the energy evaporates water from the oceans, and a small fraction powers the photosynthesis to grow vegetation on the planet. The energy that is absorbed by the surface and the atmosphere is reradiated back into space during the nighttime. The radiation (wavelengths) from any source depends on the temperature of the source. Since the sun is at a very high temperature, the radiation from the sun includes visible, non-visible infrared, and some ultraviolet radiation (short waves). However, the radiation going into space from the Earth is from a low temperature surface, and it is in the form of infrared radiation (non visible!). The atmosphere—in addition to oxygen and nitrogen—contains gases such as argon, carbon dioxide, water vapor, methane (CH_4), and laughing gas (N_2O). Some of these gases have the ability to absorb the infrared radiation, which is going out

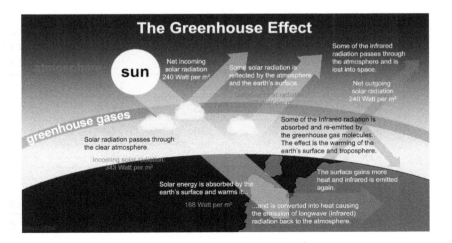

Figure 4.8

Illustration of greenhouse effect.

into space. These gases are called greenhouse gases. This natural greenhouse effect—absorption of a part of outgoing infrared radiation—keeps the average temperature of the Earth (including summer, winter, night and day, and all latitudes) at 59°F. If there were no greenhouse gases in the atmosphere, the radiation received during the daytime would be escaping in the night and the Earth would be at an average temperature (including summer, winter, night and day, and all latitudes) of 0°F. At this temperature, nothing grows on this planet and life cannot be supported.

So is the greenhouse effect bad? The answer is a resounding NO! However, the concentration of greenhouse gases in the atmosphere has been changing over the past 150 years. Table 4.2 shows the change in the concentration of these greenhouse gases. The table lists some of the main greenhouse gases and their concentrations in pre-industrial times and in 1994; atmospheric lifetimes; anthropogenic sources; and Global Warming Potential (GWP). GWP is an index defined as the cumulative radiative forcing (infrared radiation absorption) between the present and some chosen time horizon caused by a unit mass of gas emitted now, expressed relative to a reference gas such as CO_2, as is used here. GWP is an attempt to provide a simple measure of the relative radiative effects of different greenhouse gases.

Since pre-industrial times, atmospheric concentrations of CO_2, CH_4, and N_2O have climbed by over 31 percent, 151 percent, and 17 percent, respectively. Scientists have confirmed that this is primarily due to human activity. Burning coal, oil, and gas, and cutting down forests are largely responsible. The sources for emission of these gases are listed in Table 4.3.

Table 4.2   *Change in the Atmospheric Concentration of Greenhouse Gases*

GREENHOUSE GASES	PRE-INDUSTRIAL CONCENTRATION (PPBV)	CONCENTRATION (2005) (PPBV)	ATMOSPHERIC LIFETIME (YEARS)	GLOBAL WARMING POTENTIAL (GWP)*
Carbon dioxide (CO_2)	278,000	379,000	Variable	1
Methane (CH_4)	715	1774	12.2 ± 3	21
Nitrous oxide (N_2O)	270	319	120	310
CFC-12	0	0.503	102	6200–7100**
HCFC-22	0	0.105	12.1	
Perfluoromethane (CF_4)	0	0.07	50,000	6,500
Sulfur hexa-floride (SF_6)	0	0.032	3,200	23,900

*GWP is for 100-year time horizon.

**Net GWP including indirect effect due to ozone depletion.

Table 4.3   *Greenhouse Gases and their Source of Emissions*

GREENHOUSE GAS	SOURCE
Carbon dioxide (CO_2)	Combustion of solid waste, fossil fuels (oil, natural gas, and coal), and wood and wood products
Chlorofluorocarbons (CFC)	Used in refrigerants, foam manufacture, chemical solvents, and halons (used in fire extinguishers)
Methane (CH_4)	Production and transport of coal, natural gas, and oil. Methane emissions also result from the decomposition of organic wastes in municipal solid waste landfills and the raising of livestock
Nitrous oxide (N_2O)	Agricultural and industrial activities, as well as during combustion of solid waste and fossil fuels
Hydrofluorocarbons (HFCs), perfluorocarbons (PFCs), and sulfur hexafluoride (SF_6)	Industrial processes

Figure 4.9 shows the distribution of emission by greenhouse gases. Energy-related CO_2 and CH_4 account for 90 percent of the total greenhouse gas emissions in the United States. This highlights the impact of energy use on the environment. The actual measurement data from Mauna Loa Observatory show that the concentration of CO_2 in the air was 379 ppm in 2005. Figure 4.10 shows a plot of CO_2 concentration in the atmosphere since 1958. An increase in the CO_2 concentration from a pre-industrial level of 318 ppm (parts per million parts) in 1958 to about 379 ppm in the year 2005 can be seen from Figure 4.10.

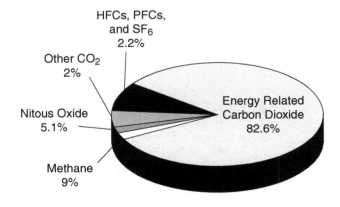

Figure 4.9   *Distribution of emissions by greenhouse gases.*

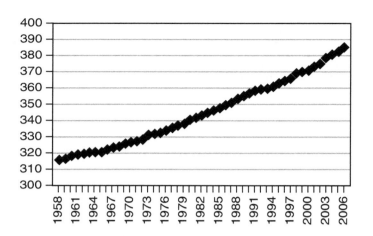

Figure 4.10

Change in the global concentration of CO_2.

Ice in the polar regions traps the air of that time period. New ice is deposited over the previously deposited ice, trapping the air from the past. Thus, analysis of ice cores provides the composition of past air, which can be used to determine past temperatures. It is evident that the rapid increase in CO_2 concentrations has been occurring since the onset of industrialization. This increase in greenhouse gases is believed to be causing an increase in the global temperature over the past 150 years, as shown in Figure 4.11. The mean increase in the global temperature over the past one century is about 1°F.

The increase has closely followed the increase in CO_2 emissions from fossil fuels. The main question is: Did the atmospheric CO_2 concentration and temperatures change prior to the pre-industrial periods?

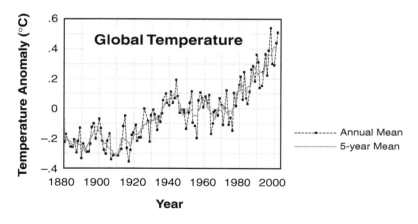

Figure 4.11

Global annual surface temperature increase over the past 120 years.
SOURCE: *http://www.giss.nasa.gov/research/observe/surftemp/2002fig1.gif*

The ice core samples indicated that the atmospheric concentration of CO_2 and temperatures fluctuated in the past significantly as shown in Figure 4.12.

It is very interesting to note that the average global temperature fluctuated in the past, increasing in the past up to about 4°F and also decreasing by about 17°F during various cycles when human activity was absent.

"More than six billion people live on this planet now who were not there during those earlier natural temperature cycles, and any chances that jeopardize the existence of this humankind must be taken seriously."

The main question is: How can one attribute a mean global temperature increase of 1°F over the past century to human activities when the past global temperature fluctuated from +4°F to –17°F from the current mean temperature when there was no human intervention? Some argue that temperature change is natural and since it is cyclical we, the humans, might not be influencing the current change in the climate. The most important difference between now and then that we have to keep in mind is the human species. More than six billion people live on this planet now who were not there during those earlier natural temperature cycles, and any chances that jeopardize the existence of this humankind must be taken seriously.

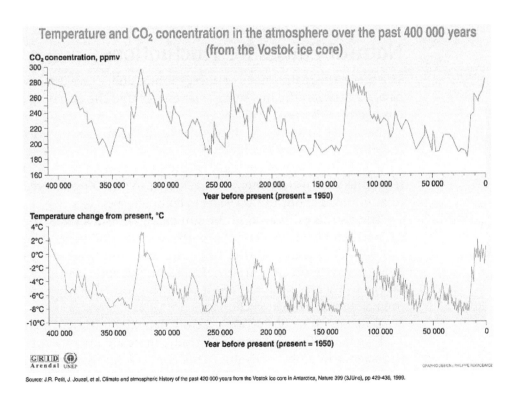

Temperature and CO_2 concentration in the atmosphere over the past 400 000 years (from the Vostok ice core)

Source: J.R. Petit, J. Jouzel, et al. Climate and atmospheric history of the past 420 000 years from the Vostok ice core in Antarctica, Nature 399 (3JUne), pp 429-436, 1999.

| Figure 4.12 | *Global temperature and CO_2 concentration changes over the past 400,000 years.* |

SOURCE: *UNDP, http://www.grida.no/climate/vital/02.htm*
Reproduced with permission from UNDP/GRID Arendal.

Reasons for Concern

Figure 4.12 shows that the temperature fluctuated significantly. Those fluctuations lead to the glacial and inter-glacial ice ages. The temperature increase over the last 150 years is not significant compared to the changes in past known history. However, a closer examination of the CO_2 profile suggests that the CO_2 concentration and the temperature correlate reasonably well. The CO_2 concentration did not rise above 310 ppm at any time in the past 400,000 years although the temperature did. Therefore, it can be concluded that this increase is something that the atmosphere did not experience earlier. It can be attributed to human activity. When 310 ppm of CO_2 during the prehistoric period could increase the temperature by about 4°F, what is the impact of the current CO_2 concentration of 379 ppm in 2005? How long does it take for the temperature to come to a level that really corresponds to this concentration? Or are we already there? And what will be the consequences if the concentration of CO_2 increases to 550 ppm (double the pre-industrial concentration), or even to 700 ppm (both of which are proposed to be likely scenarios with increased fossil fuel consumption)? If the temperature already reached the maximum temperature corresponding to 379 ppm, then some other factors are cooling the planet that were absent in the previous cycles. Let's examine the reasons or causes for the natural fluctuations.

Natural Causes for Fluctuations

The sun is the main source of energy input, and, as we discussed earlier, the net balance between the incoming solar energy and the outgoing energy causes the temperature changes.

Earth is continuously moving around the sun, and, based on its position, the incoming energy changes.

The Earth's axis of rotation is tilted at an angle of 23.5°, and this tilt goes from one side to the other and back over in 40,000-year cycles. Earth's axis of rotation takes about 21,000 years to complete a cycle.

The Earth's orbit around the sun changes from a circular path to an elliptical path and back to a circular path over 100,000 years. These are long-term changes. On a much shorter term, the radiation from the sun can be affected by the activity on the surface of the sun. Sunspots (intense flares on the surface) can increase the radiation from the sun. The increase in solar activity occurs over an 11-year cycle.

What are some of the other factors that are limiting the temperature?

Like many fields of scientific study, there are uncertainties associated with the science of global warming. This does not imply that all things are equally uncertain. Some aspects of the science are based on well-known physical laws and documented trends, while other aspects range from "near certainty" to "big unknowns."

Most scientists have pretty much agreed that human activities are changing the composition of Earth's atmosphere. Increasing levels of greenhouse gases like carbon dioxide (CO_2) in the atmosphere since pre-industrial times have been well documented. This atmospheric buildup of carbon dioxide and other greenhouse gases is largely the result of human activities.

It is well accepted by scientists that greenhouse gases trap heat in the Earth's atmosphere and tend to warm the planet. By increasing the levels of greenhouse gases in the atmosphere, human activities are augmenting Earth's natural greenhouse effect. The key greenhouse gases emitted by human activities remain in the atmosphere for periods ranging from decades to centuries.

A warming trend of about 1°F has been recorded since the late 19th century. Warming has occurred in both the northern and southern hemispheres, and over the oceans. Confirmation of 20th-century global warming is further substantiated by melting glaciers, decreased snow cover in the northern hemisphere, and even warming below ground.

Determining to what extent the human-induced accumulation of greenhouse gases since pre-industrial times is responsible for the global warming trend is not easy. This is because other factors, both natural and human, affect our planet's temperature. Scientific understanding of these other factors—most notably natural climatic variations, changes in the sun's energy, and the cooling effects of pollutant aerosols—remains incomplete or uncertain. Figure 4.13 illustrates the cooling effects of atmospheric particles and aerosols.

The cooling factors

White upper side of clouds

Volcanic eruptions

biomass burning (forest fires)

Burning of coal and oil

Ice and snow

AEROSOLS

Deserts, and dust from sandstorms

AEROSOLS (Sulphates)

Barren lands

AEROSOLS

Energy reflected

Albedo: ability of a surface to reflect light.

Aerosols: tiny particles of liquid or dust suspended in the atmosphere (most important anthropogenic aerosols is sulphate produced from SO_2)

GRID Arendal
GRAPHIC DESIGN : PHILIPPE REKACEWICZ

Sources: Radiative forcing of climate change, the 1994 report of the scientific assessment working group of IPCC, summary for policymakers, WMO, UNEP; L.D. Danny Harvey, Climate and global environmental change, Prentice Hall, pearson Education, Harlow, United Kingdom, 2000.

Figure 4.13   *Cooling effects of atmospheric particles such as sulfates, and the climate's response to changes in the atmosphere.*

SOURCE: *United Nations Development Program/GRID—Arendal (reproduced with permission).*

Nevertheless, the Intergovernmental Panel on Climate Change (IPCC) stated there was a "discernible" human influence on climate and that the observed warming trend is "unlikely to be entirely natural in origin." In the most recent Fourth Assessment Report (2007), IPCC wrote, *"The primary source of the increased atmospheric concentration of carbon dioxide since the pre-industrial period results from fossil fuel use, with land use change providing another significant but smaller contribution,"* and *"Warming of the climate system is unequivocal, as is now evident from observations of increases in global average air and ocean temperatures, widespread melting of snow and ice, and rising global average sea level."*

Computer Models and Feedbacks

Computer models solve complex mathematical equations used to describe various processes involved in climate change through well defined physical laws. However, simplifications have to be made due to either insufficient knowledge of a certain process or due to high computational requirements. Several factors can affect the predictions.

Clouds and their Influence

More clouds will form as a result of evaporation of water due to global warming. Whether a given cloud will heat or cool the surface depends on several factors, including the cloud's height, its size, and the makeup of the particles that form the cloud. Low, thick clouds primarily reflect solar radiation and cool the surface of the Earth. This will negate global warming. This is called negative feedback. High, thin clouds primarily transmit incoming solar radiation; at the same time, they trap some of the outgoing infrared radiation emitted by the Earth and radiate it back downward, thereby warming the surface of the Earth. This is likely to increase warming, and it is called positive feedback.

The balance between the cooling and warming actions of clouds is very close although, overall, cooling predominates. Imagine the difficulty in predicting cloud height and thickness decades into the future.

Fine Particles (Aerosols) in the Atmosphere

The amount of fine particles or aerosols in the air has a direct effect on the amount of solar radiation hitting the Earth's surface. Aerosols may have significant local or regional impact on temperature. Atmospheric factors shown in Figure 4.14 include natural factors such as clouds, volcanic eruptions, natural biomass (forest) burning, and dust from storms. In addition, human-induced factors such as biomass burning (forest and agricultural fires) and sulfate aerosols from burning coal contribute tiny particles that contribute to cooling. When Mount Pinatubo erupted in the Philippines in 1991, an estimated 20 million tons of sulfur dioxide and ash particles blasted more than 12 miles high into the atmosphere. The eruption caused widespread destruction and human casualities. Gases and solids injected into the stratosphere circled the globe for three

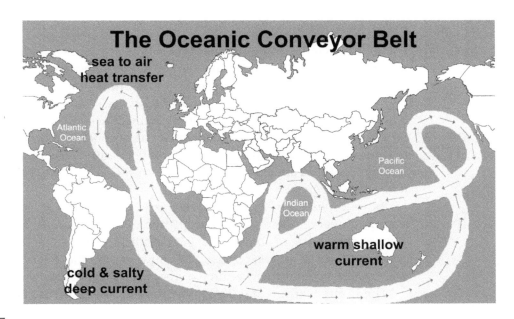

Figure 4.14

Transport of heat through the Great Ocean Conveyor.

weeks. Volcanic eruptions of this magnitude can impact global climate, reducing the amount of solar radiation reaching the Earth's surface, lowering temperatures in the troposphere, and changing atmospheric circulation patterns. This is a negative feedback. The extent to which this occurs is an uncertainty. Water vapor is a greenhouse gas, but at the same time the upper white surface of clouds reflects solar radiation back into space. Albedo—reflections of solar radiation from surfaces on the Earth—creates difficulties in exact calculations. If, for example, the polar icecap melts, the albedo will be significantly reduced. Open water absorbs heat, while white ice and snow reflect it.

Oceans

Oceans play a vital role in the energy balance of the Earth, and they are also a wild card in climate change. It is known that the top 10 feet of the oceans can hold as much heat as the entire atmosphere above the surface. However, most of the incoming energy is incident on the equatorial region. The water in the oceans in these regions is warmer and needs to be transported to the northern latitudes. This occurs due to natural variations in the temperatures of the water from prevailing winds that cause disturbances in the surface waters. The cold, dense water from the polar regions flows into and sinks to the bottom of the oceans, pushing the warm waters from the surface towards the northern latitudes. This, along with the winds, forms an ocean conveyor belt carrying the surface heat to the northern latitudes. In the process, the water slowly cools and sinks to the bottom. This is also a way to carry the dissolved carbon dioxide in the surface water into the deep oceans. This is shown in the Figure 4.14.

Polar Ice Caps

Polar ice caps reflect the sunlight and cool the atmosphere. When the polar ice caps melt due to global warming, darker surfaces underneath will be exposed; darker surfaces absorb more solar energy thus warming the atmosphere even more.

With increased understanding of the processes and increased computing power, predictive capabilities are getting refined. According to the recent (2007) IPCC report, scientists from 130 countries have concluded that "it is very likely that human activity is the cause for climate change." The report also predicts that temperatures are most likely to rise by 1.8°C–4°C by 2100. But the possible range is much greater; 1.1°C–6.4°C.

As atmospheric levels of greenhouse gases continue to rise, scientists estimate average global temperatures will continue to rise as a result. For the next two decades a warming of about 0.2°C per decade is projected by IPCC. Even if the concentrations of all greenhouse gases and aerosols had been kept constant at year 2000 levels, a further warming of about 0.1°C per decade would be expected.

The Intergovernmental Panel on Climate Change (IPCC) states that even the low end of this warming projection "would probably be greater than any seen in the last 10,000 years, but the actual annual to decadal changes would include considerable natural variability."

Impacts of Global Warming

Health

It is well known that extremely hot temperatures increase the number of deaths. People with cardiovascular diseases are particularly vulnerable because the heart must work harder to keep the body cool during hot weather. Heat exhaustion, dehydration, and some respiratory illnesses are prevalent in hot weather. Higher air temperatures also increase the concentration of ozone at ground level (as we will see later). Because of the warmer temperatures diseases such as malaria, dengue fever, yellow fever, and encephalitis that are spread by mosquitoes and other insects could become prevalent once again in northern latitudes where these are thought to have been eradicated.

Water Resources

Changing climate is expected to increase both evaporation and precipitation in most areas of the United States. In those areas where evaporation increases more than precipitation, soil will become drier, lake levels will drop, and rivers will carry less water. Lower river flows and lower lake levels could impair navigation, hydroelectric power generation, and water quality, and reduce the supplies of water available for agricultural, residential, and industrial uses. Some areas may experience both increased flooding during winter and spring, as well as lower supplies during summer.

Polar Regions

Climate models indicate that global warming will be felt most acutely at high latitudes, especially in the Arctic where reductions in sea ice and snow cover are expected to lead to the greatest relative temperature increases. Ice and snow cool the climate by reflecting solar energy back to space, so a reduction in their extent would lead to greater warming in the region.

Forests

The projected 2°C (3.6°F) warming could shift the ideal range for many North American forest species by about 300 km (200 mi.) to the north. If the climate changes slowly enough, warmer temperatures may enable the trees to colonize north into areas that are currently too cold, at about the same rate as southern areas became too hot and dry for the species to survive. If the Earth warms 2°C (3.6°F) in 100 years, however, the species would have to migrate about 2 miles every year. Poor soils may also limit the rate at which tree species can spread north. Several other impacts associated with changing climate further complicate the picture. On the positive side, CO_2 has a beneficial fertilization effect on plants, and it also enables plants to use water more efficiently. These effects might enable some species to resist the adverse effects of warmer temperatures or drier soils. On the negative side, forest fires are likely to become more frequent and severe if soils become drier.

Coastal Zones

Sea level is rising more rapidly along the U.S. coast than worldwide. Studies by EPA and others have estimated that along the Gulf and Atlantic coasts, a one-foot (30 cm) rise in sea level is likely by 2050. In the next century, a two-foot rise is most likely, but a four-foot rise is possible. Rising sea level inundates wetlands and other low-lying lands, erodes beaches, intensifies flooding, and increases the salinity of rivers, bays, and groundwater tables. Low-lying countries like Maldives located in the Indian Ocean and Bangladesh may be severely affected. The world may see global warming refugees from these impacts.

Scientists have identified that our health, agriculture, water resources, forests, wildlife, and coastal areas are vulnerable to the changes that global warming may bring. But projecting what the exact impacts will be over the 21st century remains very difficult. This is especially true when one asks how a local region will be affected.

Table 4.3 *Predicted Effects of Climate Change and their Probability of Occurrence*

EFFECT	PROBABILITY OF OCCURRENCE
Warmer and fewer cold days and nights over most land areas	Virtually certain
Warmer and more frequent hot days and nights over most land areas	Virtually certain
Warm spells/heat waves; frequency increases over most land areas	Very likely
Heavy precipitation events	
Frequency (or proportion of total rainfall from heavy falls) increases over most areas	Very likely
Area affected by droughts increases	Very likely
Increased incidence of extreme high sea level	Likely

IPCC definitions of probability of occurrence:

Virtually certain:	more than 99%	Very likely:	more than 90%
Extremely likely:	more than 95%	Likely:	more than 66%

Scientists are more confident about their projections for large-scale areas (e.g., global temperature and precipitation change, and average sea level rise) and less confident about the ones for small-scale areas (e.g., local temperature and precipitation changes, altered weather patterns, and soil moisture changes). This is largely because the computer models used to forecast global climate change are still ill-equipped to simulate how things may change at smaller scales. Table 4.3 lists the possible effects and the probability of occurrence as reported by the IPCC report (2007).

Is There a Solution for This Potential Global Warming?

Today there is no single agreed-upon solution because governments of several countries are still not clear about the path. There is certainty that human activities are rapidly adding greenhouse gases to the atmosphere and that these gases tend to warm our planet. This is the basis for concern about global warming. Global warming poses real risks. The exact nature of these risks remains uncertain to some degree.

The Fourth IPCC report (2007) "clearly assesses the impacts of climate change in different parts of the world and we have far greater regional detail than in [our previous global assessment in] 2001 on things like glacier melting, and what the implications of that melting will be; on sea level rise, which clearly threatens a number of countries in the world including mega-deltas which are particularly vulnerable; and on agriculture, which has implications for food security."

Kyoto Protocol to the United Nations Framework Convention on Climate Change

The main purpose of this protocol is to limit or reduce greenhouse gas emissions. The countries that are parties to this protocol are Annex 1 countries. The Parties included in Annex I shall, individually or jointly, ensure that their aggregate anthropogenic carbon dioxide equivalent emissions of the greenhouse gases do not exceed their assigned amounts, calculated pursuant to their quantified emission limitation and reduction commitments inscribed with a view to reducing their overall emissions of such gases by at least 5 percent below 1990 levels in the commitment period 2008 to 2012. Only Parties to the Convention that have also become Parties to the Protocol (i.e, by ratifying, accepting, approving, or acceding to it) will be bound by the Protocol's commitments. To date, 175 Parties have ratified the Protocol. Of these, 36 countries and the EEC are required to reduce greenhouse gas emissions below levels specified for each of them in the treaty. For the Protocol to come fully into force, the pact needed to be ratified by countries accounting for at least 55 percent of 1990 carbon dioxide emissions. With countries like the U.S. and Australia unwilling to join the pact, the key to ratification came when Russia, which accounted for 17 percent of 1990 emissions, signed onto the agreement on November 5, 2004. During these seven years that it took for the agreement to be in full force, greenhouse gas emissions for several countries have increased significantly and have fallen behind the schedule in meeting the emission limits. While the governments are debating on possible solutions and policies, individuals can solve the problem to some extent by their own decisions.

Global Warming: Your "Power" in the Environmental Protection

What difference can an individual make? When faced with this question, individuals should recognize that collectively they can make a difference. In some cases, it only takes a little change in lifestyle and behavior to make some big changes in greenhouse gas reductions. For other types of actions, the changes are more significant. When that action is multiplied by the 300 million people in the U.S. or the 6.5 billion people worldwide, the savings are significant.

Energy efficiency means doing the same (or more) with less energy.

When individual action is multiplied by the 300 million people in the U.S. or the 6.5 billion people worldwide, the savings are significant.

How can you contribute to the solution?

- ☞ Cut your utility bills by purchasing energy-efficient appliances, fixtures, and other home equipment and products. The average house is responsible for more air pollution and carbon dioxide emissions than the average car.
  - You can reduce your energy consumption by up to 30 percent by purchasing home products—appliances, new homes computers, copiers, fax machines—that display the ENERGY STAR® label.
  - When your lights burn out, replace them with energy-efficient compact fluorescent lights.
  - Insulate your home and tune up your furnace.
- ☞ Save on water use in your home.
  - Use low-flow faucets in your showers and sinks.
  - Replace toilets with water-saving lavatories.
  - Lower the temperature on your hot water tank to 120 or even 110°F.
  - Insulate your water heater and all water pipes to reduce heat loss.
- ☞ When you remodel, build, or buy a new home, incorporate all of these energy efficiency measures—and others.
- ☞ Purchase "Green Power" for your home's electricity if available from your utility.
  - Green power is electricity that is generated from renewable sources such as solar, wind, geothermal, or biomass. Although the cost may be slightly higher, you'll know that you're buying power from an environmentally friendly energy source.
  - Use locally grown or manufactured products.

We will discuss why and how these steps reduce emissions in later chapters.

> Each of us, in the U.S., contributes about 19 tons of carbon dioxide emissions per year, whereas the world average per capita is about 6 tons.
>
> The good news is that there are many ways you can help reduce carbon dioxide pollution and improve the environment for you and your children.

Acid Rain

Acid rain is a serious environmental problem around the world, particularly in Asia and Europe. It also affects large parts of the U.S. and Canada. Acidic pollutants such as SO_2 and NO_x are emitted into the environment by combustion of fossil fuels. Most of the sulfur in any fuel combines with oxygen and forms SO_2 in the combustion chamber. This SO_2 when emitted into the atmosphere slowly oxidizes to SO_3. SO_3 is readily soluble in water in the clouds and forms H_2SO_4 (sulfuric acid).

$$S + O_2 \rightarrow SO_2 + \tfrac{1}{2} O_2 \text{ (in the atmosphere)} \rightarrow SO_3 + H_2O \rightarrow H_2SO_4 \text{ (sulfuric acid)}$$

Most of the NO_x that is emitted is in the form of NO. This NO is oxidized in the atmosphere to NO_2. NO_2 is soluble in water and forms HNO_3 (nitric acid).

$$NO + O_2 \text{ in the atmosphere} \rightarrow NO_2$$

$$2NO_2 + H_2O \rightarrow HNO_2 \text{ (nitrous acid)} + HNO_3 \text{ (nitric acid)}$$

Sunlight increases the rate of most of these reactions. The result is a mild solution of sulfuric acid and nitric acid. "Acid rain" is a broad term used to describe several ways that acids fall out of the atmosphere. A more precise term is acid deposition, which has two parts: wet and dry. Wet deposition refers to acidic rain, fog, and snow. As this acidic water flows over and through the ground, it affects a variety of plants and animals. The strength of the effects depends on many factors, including how acidic the water is, the chemistry and buffering capacity of the soils involved, and the types of fish, trees, and other living things that rely on the water. Dry deposition refers to acidic gases and particles. About half of the acidity in the atmosphere falls back to earth through dry deposition. The wind blows these acidic particles and gases onto buildings, cars, homes, and trees. Dry deposited gases and particles can also be washed from trees and other surfaces by rainstorms. When that happens, the runoff water adds those acids to the acid rain, making the combination more acidic than the falling rain alone.

Prevailing winds blow the compounds that cause both wet and dry acid deposition across state and national borders, and sometimes over hundreds of miles. Figure 4.15 shows the process of acid deposition.

How Is Acid Rain Measured?

Acid rain is measured using a scale called pH. The lower a substance's pH, the more acidic it is. The pH scale is a measure of hydrogen ion concentration measured as a negative logarithm. It measures how acidic or basic a substance is. As

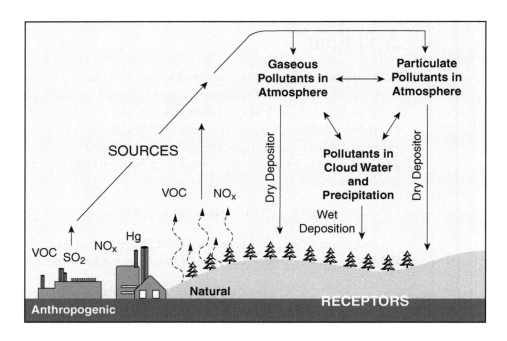

Figure 4.15

Formation and deposition of acid rain.
SOURCE: *http://www.epa.gov/airmarkets/acidrain/index.html#what*

shown in Figure 4.16, pH ranges from 0 to 14. A pH of 7 is neutral. A pH less than 7 is acidic, and a pH greater than 7 is basic. Each whole pH value below 7 is ten times more acidic than the next higher value. For example, a pH of 4 is ten times more acidic than a pH of 5 and 100 times (10 times 10) more acidic than a pH of 6. The same holds true for pH values above 7, each of which is ten times

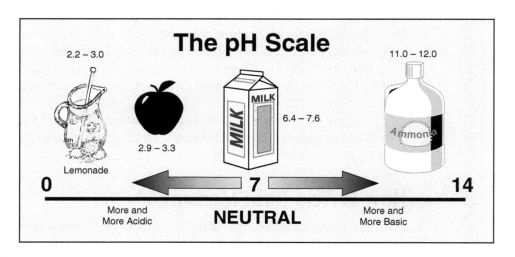

Figure 4.16

Acidity scale.
SOURCE: *http://www.epa.gov/airmarkets/acidrain/ph.html*

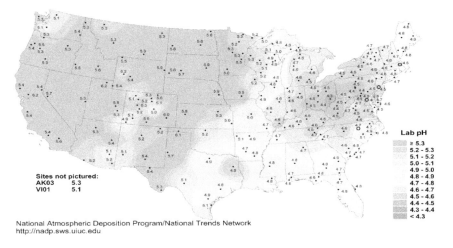

Figure 4.17

pH measurements of surface water in the U.S. (2005).

more alkaline (another way to say basic) than the next lower whole value. For example, a pH of 10 is ten times more alkaline than a pH of 9. Pure water has a pH of 7.0. Normal rain is slightly acidic because carbon dioxide dissolves into it, so it has a pH of about 5.5. As of the year 2005 (Figure 4.17), the most acidic rain falling in the U.S. has a pH of about 4.4–4.5.

Effect of Acid Rain on Health and the Environment

The effects of acid deposition are illustrated in Figure 4.18. Acid rain causes acidification of lakes and streams and contributes to damage of trees at high elevations and to many sensitive forest soils. In addition, acid rain accelerates the decay of building materials and paints, including irreplaceable buildings, statues, and sculptures that are part of our nation's cultural heritage. Prior to falling to the earth, SO_2 and NO_x gases and their particulate matter derivatives, sulfates and nitrates, contribute to visibility degradation and harm public health.

Surface Waters (e.g., Lakes and Streams) and Animals Living in Them

Acid rain causes acidification of lakes and streams and contributes to the damage of trees at high elevations (for example, red spruce trees above 2,000 feet) and many sensitive forest soils. Several regions in the U.S. were identified as containing many of the surface waters sensitive to acidification. They include

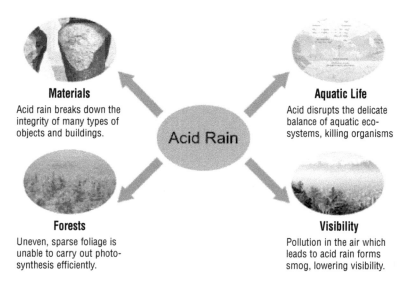

Figure 4.18

Illustrations of the effects of acid deposition.

the Adirondacks and Catskill Mountains in New York State, the mid-Appalachian highlands along the east coast, the upper Midwest, and mountainous areas of the Western United States. Some types of plants and animals can tolerate acidic waters. Others, however, are acid-sensitive and will be lost as the pH declines. Figure 4.19 shows that not all fish, shellfish, or the insects that they eat can tolerate the same amount of acid; for example, frogs can tolerate water that is more acidic (has lower pH) than trout.

Forests

Acid rain does not usually kill trees directly. Instead, it is more likely to weaken trees by damaging their leaves, limiting the nutrients available to them, or exposing them to toxic substances slowly released from the soil. Quite often, injury

Species	pH 6.5	pH 6.0	pH 5.5	pH 5.0	pH 4.5	pH 4.0
Trout						
Bass						
Perch						
Frogs						
Salamanders						
Clams						
Crayfish						
Snails						
Mayfly						

Figure 4.19

Effect of acid rain on aquatic life.

DATA SOURCE: *http://www.epa.gov/airmarkets/acidrain/surfacewater.html*

or the death of trees is a result of these effects of acid rain in combination with one or more additional threats.

Scientists know that acidic water dissolves the nutrients and helpful minerals in the soil and then washes them away before trees and other plants can use them to grow. At the same time, acid rain causes the release of substances that are toxic to trees and plants, such as aluminum, into the soil. Scientists believe that this combination of the loss of soil nutrients and the increase of toxic aluminum may be one way that acid rain harms trees.

Such substances also wash away in the runoff and are carried into streams, rivers, and lakes. More of these substances are released from the soil when the rainfall is more acidic.

However, trees can be damaged by acid rain even if the soil is well buffered. Forests in high mountain regions often are exposed to greater amounts of acid than other forests because they tend to be surrounded by acidic clouds and fog that are more acidic than rainfall. Scientists believe that essential nutrients are stripped away when leaves and needles are frequently bathed in this acid fog. This loss of nutrients in their foliage makes trees more susceptible to damage by other environmental factors, particularly cold winter weather.

Materials

Acid rain and the dry deposition of acidic particles contribute to the corrosion of metals (such as bronze) and the deterioration of paint and stone (such as marble and limestone). These effects seriously reduce the value to society of buildings, bridges, cultural objects (such as statues, monuments, and tombstones), and cars. Most of the monumental buildings are made of marble, which is composed of calcite (calcium carbonate, $CaCO_3$). When marble is exposed to acid the following reaction takes place. Calcium carbonate turns into calcium sulfate ($CaSO_4$) and releases CO_2 and H_2O.

$$CaCO_3 + H_2SO_4 \rightarrow CaSO_4 + CO_2 + H_2O$$

While $CaCO_3$ is practically insoluble in water, $CaSO_4$ is very soluble in water, thus making the statues, monuments, and tombstones dissolve in rain.

Visibility

Sulfates and nitrates that form in the atmosphere from sulfur dioxide (SO_2) and nitrogen oxides (NO_x) emissions contribute to visibility impairment, meaning we can't see as far or as clearly through the air. Sulfate particles account for 50 to 70 percent of the visibility reduction in the eastern part of the United States. This affects our enjoyment of national parks, such as the Shenandoah and the Great Smoky Mountains. Based on a study of the value national park visitors place on visibility, the visual range improvements expected at national parks of the eastern United States due to the SO_2 reductions will be worth over a billion dollars annually by the year 2010. In the western part of the United States, nitrates and carbon also play roles, but sulfates have been implicated as an important source of visibility impairment in many of the Colorado River Plateau national parks, including the Grand Canyon, Canyon lands, and Bryce Canyon.

Human Health

Acid rain looks, feels, and tastes just like clean rain. The harm to people from acid rain is not direct. Walking in acid rain, or even swimming in an acid lake, is no more dangerous than walking or swimming in clean water. However, the pollutants that cause acid rain (sulfur dioxide (SO_2) and nitrogen oxides (NO_x)) also damage human health. These gases interact in the atmosphere to form fine sulfate and nitrate particles that can be transported long distances by winds and inhaled deep into people's lungs. Fine particles can also penetrate indoors. Many scientific studies have identified a relationship between elevated levels of fine particles and increased illness and premature death from heart and lung disorders, such as asthma and bronchitis.

A decrease in nitrogen oxide emissions is also expected to have a beneficial impact on human health by reducing the nitrogen oxides available to react with volatile organic compounds and to form ozone. The impact of ozone on human health includes a number of morbidity and mortality risks associated with lung inflammation, including asthma and emphysema.

Acid Rain: Your "Power" in the Environmental Protection

- Turn off lights, computers, and other appliances when you're not using them.

- Use energy efficient appliances: lighting, air conditioners, heaters, refrigerators, washing machines, etc.

- Only use electric appliances when you need them.

- Keep your thermostat at 68°F in the winter and 72°F in the summer. You can turn it even lower in the winter and higher in the summer when you are away from home.

- Insulate your home as best you can.

- Carpool, use public transportation, or better yet, walk or bicycle whenever possible.

- Buy vehicles with low NO_x emissions and maintain all vehicles well.

Ozone and Environment

Ozone (O_3) is a tri-atomic oxygen molecule gas that occurs both in the Earth's upper atmosphere and at ground level. Ozone can be good or bad, depending on where it is found. It is a bluish gas that is harmful to breathe. Therefore, it is bad at the ground level.

Good Ozone

Ozone occurs naturally in the Earth's stratosphere (upper atmosphere)—10 to 30 miles above the Earth's surface—where it shields us from the sun's harmful ultraviolet rays, called the UVB band. Ninety percent of the Earth's ozone is in the stratosphere and is referred to as the "Ozone Layer."

Bad Ozone

In the Earth's lower atmosphere, near ground level, ozone is formed when pollutants emitted by cars, power plants, industrial boilers, refineries, chemical plants, and other sources react chemically in the presence of sunlight. Ozone pollution is a concern during the summer months when the weather conditions needed to form ground-level ozone—lots of sun and hot temperatures—naturally occur.

Good Ozone or Ozone Layer Destruction

Figure 4.20 shows the process of ozone depletion. It occurs in steps as follows:

1. CFCs and other ozone-depleting substances (ODS) are emitted into the atmosphere due to human activity.
2. CFCs are extremely stable and remain in the atmosphere for years. When the ODS molecules reach the stratosphere, about 10 miles above the Earth's surface, the UV rays break apart the ODS molecule.
3. CFCs, HCFCs, carbon tetrachloride, methyl chloroform, and other gases release chlorine atoms and halons (bromo chloro carbons), and methyl bromide releases bromine atoms. These broken halogen atoms destroy the ozone by converting into oxygen.
4. It is estimated that one chlorine atom can destroy over 100,000 ozone molecules before it is finally being removed from the stratosphere.

Ozone is constantly produced and destroyed in a natural cycle, as shown in Figure 4.21. However, the overall amount of ozone is essentially stable. This balance can be thought of as a stream's depth at a particular location. Although

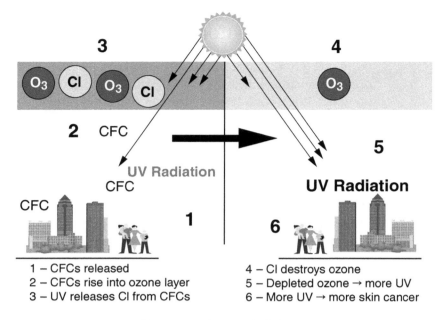

Figure 4.20

The process of ozone depletion.

SOURCE: *http://www.epa.gov/ozone/science/process.html*

1 – CFCs released
2 – CFCs rise into ozone layer
3 – UV releases Cl from CFCs

4 – Cl destroys ozone
5 – Depleted ozone → more UV
6 – More UV → more skin cancer

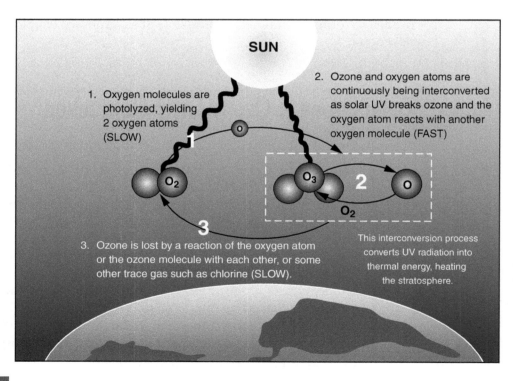

Figure 4.21

Natural production and destruction of ozone in the stratosphere.

SOURCE: *http://www.epa.gov/ozone/science/process.html*

individual water molecules are moving past the observer, the total depth remains constant. Similarly, while ozone production and destruction are balanced, ozone levels remain stable. This was the situation until the past several decades.

Large increases in stratospheric chlorine and bromine, however, have upset this balance. In effect, they have added a siphon downstream, removing ozone faster than natural ozone creation reactions can keep up. Therefore, ozone levels fall.

Since ozone filters out harmful UVB radiation, less ozone means higher UVB levels at the surface. The more depletion, the larger the increase in incoming UVB radiation. UVB has been linked to skin cancer, cataracts, damage to materials like plastics, and harm to certain crops and marine organisms. Although some UVB reaches the surface even without ozone depletion, its harmful effects will increase as a result of this problem. Ozone-Depleting Substance(s) (ODS) are CFCs, HCFCs, (used in the energy related refrigeration and air conditioning in homes, commercial buildings and cars and manufacture of foam products), halons (used in fire extinguishers), methyl bromide, carbon tetrachloride, and methyl chloroform, which are used as solvents in chemical industries.

How Much Ozone Is Lost?

Recent studies by NASA and others have indicated that about 40 percent of the ozone in Antarctica has been destroyed, and about 7 percent of the ozone is destroyed from the Arctic Circle. The area of destruction of ozone is also called the "Ozone Hole." Ozone "hole" does not mean that there is no ozone in the region. The ozone hole is defined as the area having less than 220 dobson units (DU) of ozone (concentration) in the overhead column (i.e., between the ground and space). Figure 4.22 shows the reduction in ozone concentration over Antarctica. Unfortunately, this hole in Antarctica is allowing more Australians to be exposed to UV radiation. However, if this kind of ozone destruction ever takes place in the Arctic zone, more humans (in the Northern hemisphere) would be exposed to higher levels of UVB radiation.

The size of the Southern Hemisphere Ozone hole as a function of the year is shown in Figure 4.23. It also compares the size over ten years. It can be seen that the size increased year by year. But the peak is in the spring time of the Southern hemisphere.

Effects of Ozone Layer Depletion

Skin Cancer

The incidence of skin cancer in the United States has reached epidemic proportions. One in five Americans will develop skin cancer in their lifetime, and one American dies every hour from this devastating disease. It is also predicted that

EP/TOMS Version 8 Total Ozone for Sep 30, 2005

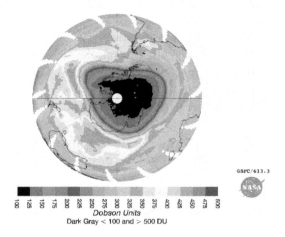

Dobson Units
Dark Gray < 100 and > 500 DU

Ozone concentration in the Antarctica.
SOURCE: *http://jwocky.gsfc.nasa.gov/multi/recent_ozone91200.gif*

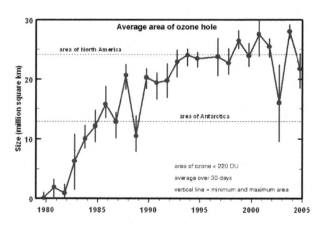

Southern hemisphere ozone hole area.
SOURCE: *http://jwocky.gsfc.nasa.gov/multi/oz_hole_area.gif*

two out of three Australians will have some form of skin cancer due to higher exposure to UV rays in the Southern Hemisphere. Medical research is helping us understand the causes and effects of skin cancer. Many health and education groups are working to reduce the incidence of this disease, of which 1.3 million cases have been predicted for 2000 alone, according to The American Cancer Society. Figure 4.24 shows the sources of ozone depleting substances.

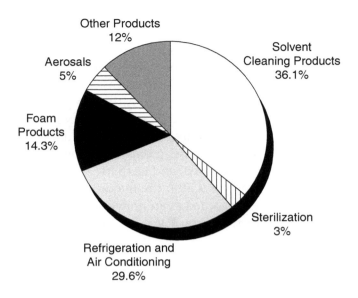

Figure 4.24

Various sources that produce ODS.

Melanoma

Melanoma, the most serious form of skin cancer, is also one of the fastest growing types of cancer in the United States. Many dermatologists believe there may be a link between childhood sunburns and melanoma later in life. Melanoma cases in this country have more than doubled in the past two decades, and the rise is expected to continue.

Non-melanoma Skin Cancers

Non-melanoma skin cancers are less deadly than melanomas. Nevertheless, left untreated, they can spread, causing disfigurement and more serious health problems. More than 1.2 million Americans will develop non-melanoma skin cancer in 2000 while more than 1,900 will die from the disease. There are two primary types of non-melanoma skin cancers. These two cancers have a cure rate as high as 95 percent if detected and treated early. The key is to watch for signs and to seek medical treatment.

Basal cell carcinomas are the most common type of skin cancer tumors. They usually appear as small, fleshy bumps or nodules on the head and neck, but they can occur on other skin areas. Basal cell carcinomas grow slowly and rarely spread to other parts of the body. They can, however, penetrate to the bone and cause considerable damage. Squamous cell carcinomas are tumors that may appear as nodules or as red, scaly patches. This cancer can develop into large masses, and, unlike basal cell carcinoma, it can spread to other parts of the body.

Other Skin Damage

Other UV-related skin disorders include actinic keratoses and premature aging of the skin. Actinic keratoses are skin growths that occur on body areas exposed to the sun. The face, hands, forearms, and the "V" of the neck are especially susceptible to this type of lesion. Although premalignant, actinic keratoses are a risk factor for squamous cell carcinoma. Look for raised, reddish, rough-textured growths and seek prompt medical attention if you discover them. Chronic exposure to the sun also causes premature aging, which over time can make the skin become thick, wrinkled, and leathery. Since it occurs gradually, often manifesting itself many years after the majority of a person's sun exposure, premature aging is often regarded as an unavoidable, normal part of growing older. With proper protection from UV radiation, however, most premature aging of the skin can be avoided.

Cataracts and Other Eye Damage

Cataracts are a form of eye damage in which a loss of transparency in the lens of the eye clouds vision. If left untreated, cataracts can lead to blindness. Research has shown that UV radiation increases the likelihood of certain cataracts. Although curable with modern eye surgery, cataracts diminish the eyesight of millions of Americans and cost billions of dollars in medical care each year. Other kinds of eye damage include pterygium (i.e., tissue growth that can block vision), skin cancer around the eyes, and degeneration of the macula (i.e., the part of the retina where visual perception is most acute). All of these problems can be lessened with proper eye protection from UV radiation.

Immune Suppression

Scientists have found that overexposure to UV radiation may suppress proper functioning of the body's immune system and the skin's natural defenses. All people, regardless of skin color, might be vulnerable to effects including impaired response to immunizations, increased sensitivity to sunlight, and reactions to certain medications.

International Action

In 1987 the Montreal Protocol, an international environmental agreement, established requirements that began the worldwide phase out of ozone-depleting CFCs (chlorofluorocarbons). These requirements were later modified, leading to the phase out in 1996 of CFC production in all developed nations.

Ozone Layer Depletion: Your "Power" in the Environmental Protection

- Make sure that technicians working on your car air conditioner, home air conditioner, or refrigerator are certified by an EPA-approved program to recover the refrigerant (this is required by law).

- Have your car and home air conditioner units and refrigerator checked for leaks. When possible, repair leaky air conditioning units before refilling them.

- Contact local authorities to properly dispose of refrigeration or air conditioning equipment.

- Protect yourself against sunburn. Minimize sun exposure during midday hours (10 A.M. to 4 P.M.). Wear sunglasses, a hat with a wide brim, and protective clothing with a tight weave. Use a broad spectrum sunscreen with a sun protection factor (SPF) of at least 15. To be safer, 30 is better.

Ground Level Ozone or "Bad Ozone" Formation and Photochemical Smog

Ozone is a secondary pollutant that forms from the primary pollutants such as Volatile Organic Compounds (Hydrocarbons) and nitrogen oxides (NO_x) in the presence of sunlight. Its formation is mainly from automobile emissions. A typical profile of pollutants in the air of major cities is well repeatable and is shown in Figure 4.25. Early in the morning, during peak traffic hours, NO and Hydrocarbons are emitted along with CO. By mid-morning, NO is slowly oxidized to NO_2. In the presence of sunlight by mid-afternoon NO_x react with VOCs to form ozone. Ozone by itself is damaging to health and also to the environment. Ozone triggers a variety of health problems even at very low levels and may cause permanent lung damage after long-term exposure. Ozone also leads to the formation of smog or haze, causing additional problems such as decrease in visibility as well as damage to plants and ecosystems.

Basic Chemistry and Sources

$$VOCs + NO_x \xrightarrow{\quad sunlight \quad} Ozone$$
$$Ozone + NO \xrightarrow{\quad sunlight \quad} Photochemical\ Smog\,(Haze)$$

As shown in Figure 4.26, motor vehicle exhaust and industrial emissions, gasoline vapors, and chemical solvents are some of the major sources of NO_x and VOC that help to form ozone. Sunlight and hot weather cause ground-level

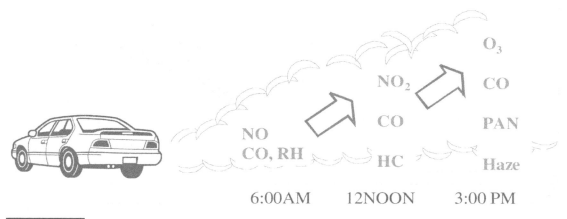

Figure 4.25

Typical pollutant profile and ozone formation during the day.

ozone to form in harmful concentrations in the air. As a result, it is known as a summertime air pollutant. Many urban areas tend to have high levels of "bad" ozone, but even rural areas are also subject to increased ozone levels because wind carries ozone and pollutants that form it hundreds of miles away from their original sources.

Health and Environmental Impact of Ground-Level Ozone

Several groups of people are particularly sensitive to ozone—especially when they are active outdoors—because physical activity causes people to breathe faster and more deeply. In general, as concentrations of ground-level ozone increase, more and more people experience health effects, the effects become more serious, and more people are admitted to the hospital for respiratory problems. When ozone levels are very high, everyone should be concerned about ozone exposure.

Health Effects of Ground Level Ozone

Ozone can irritate your respiratory system, causing you to start coughing, to feel an irritation in your throat, and/or to experience an uncomfortable sensation in your chest.

Ozone can reduce lung function and make it more difficult for you to breathe as deeply and vigorously as you normally would. When this happens, you may notice that breathing starts to feel uncomfortable. If you are exercising

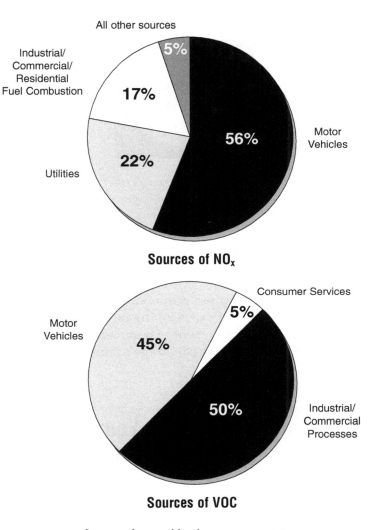

All other sources

Industrial/
Commercial/
Residential
Fuel Combustion

17%

5%

56%

Motor
Vehicles

22%

Utilities

Sources of NO$_x$

Consumer Services

Motor
Vehicles

45%

5%

50%

Industrial/
Commercial
Processes

Sources of VOC

Figure 4.26

Sources of ground level ozone precursors.

Source: *http://www.epa.gov/oar/oaqps/gooduphigh/ozone.html#7*

or working outdoors, you may notice that you are taking more rapid and shallow breaths than normal.

Ozone can aggravate asthma. When ozone levels are high, more people with asthma have attacks that require a doctor's attention or the use of additional medication. One reason this happens is that ozone makes people more sensitive to allergens, which are the most common triggers for asthma attacks. Also, asthmatics are more severely affected by the reduced lung function and irritation that ozone causes in the respiratory system.

Ozone can inflame and damage cells that line your lungs. Within a few days, the damaged cells are replaced and the old cells are shed—much in the way your skin peels after sunburn. (Figure 4.27).

Ozone can inflame the lung's lining. These photos show a healthy lung air way (left) and an inflamed lung air way (right).

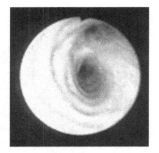 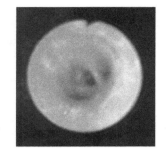

Figure 4.27

Ozone's effect on human lung.
Source: *http://epa.gov/airnow/health/smog1.html#5*

Ozone may aggravate chronic lung diseases such as emphysema and bronchitis and reduce the immune system's ability to fight off bacterial infections in the respiratory system.

Ozone may cause permanent lung damage. Repeated short-term ozone damage to children's developing lungs may lead to reduced lung function in adulthood. In adults, ozone exposure may accelerate the natural decline in lung function that occurs as part of the normal aging process.

Ground Level Ozone: Your "Power" in the Environmental Protection

- Follow gasoline refueling instructions for efficient vapor recovery. Be careful not to spill fuel and always tighten your gas cap securely.

- Keep car, boat, and other engines tuned up according to manufacturers' specification.

- Be sure your tires are properly inflated. Improper inflation can reduce gas mileage.

- Choose a cleaner commute—carpool, use public transportation, bike, or walk whenever possible.

- Use environmentally safe paints and cleaning products whenever possible.

- Some products that you use at your home or office are made with smog-forming chemicals that can evaporate into the air when you use them. Follow manufacturers' recommendations for use and properly seal cleaners, paints, and other chemicals to prevent evaporation into the air.

- Conserve electricity and set your air conditioner at a higher temperature.

- Defer use of gasoline-powered lawn and garden equipment.

- Refuel cars and trucks after dusk. During daytime the vapors produce VOCs that produce ozone and smog.

- Combine errands and reduce trips.

- Limit engine idling. Stop the engine if you have to idle more than a minute.

Sources

http://cdiac.esd.ornl.gov/trends/emis/em_cont.htm
http://www.epa.gov/globalwarming/emissions/index.html
http://www.grida.no/climate/vital/intro.htm
http://www.epa.gov/ozone/science/
http://www.epa.gov/globalwarming/climate/index.html
http://yosemite.epa.gov/oar/globalwarming.nsf/content/ActionsIndividual.html
http://www.giss.nasa.gov/edu/gwdebate/
http://www.epa.gov/air/acidrain/
http://pubs.usgs.gov/gip/acidrain/contents.html
http://www.epa.gov/air/visibility/
http://www.epa.gov/air/urbanair/6poll.html
http://www.epa.gov/air/urbanair/co/what1.html
http://www.ipcc.ch/SPM2feb07.pdf
http://ipcc-wg1.ucar.edu/wg1/Report/AR4WG1_Pub_SPM-v2.pdf
IPCC, 2007: Summary for Policymakers. In: *Climate Change 2007: The Physical Science Basis.* Contribution of Working Group I to the Fourth Assessment Report of the Intergovernmental Panel on Climate Change [S. Solomon, D. Qin, M. Manning, Z. Chen, M. Marquis, K. B. Averyt, M. Tignor and H. L. Miller (eds.)]. Cambridge University Press, Cambridge, United Kingdom and New York, NY, USA.

questions

1. What is the difference between primary and secondary pollutants?

2. How can scientists extrapolate historical climate changes by analyzing ice cores?

3. What is the greenhouse effect? Explain the difference between the greenhouse effect and global warming. Which gases contribute to the greenhouse effect?

4. Explain how ozone is formed at the ground level with the help of the basic reactions. Which end users of energy are responsible for the emissions of the compounds involved?

5. Explain how the stratospheric ozone layer is being destroyed and which sector is responsible for the emission of the gases that are responsible for this.

6. What is acid rain? How is it formed? What are the effects of acid rain?

7. List five steps that you, as an individual, can take to reduce potential global warming and explain how each of these steps will reduce emissions.

8. List five ways in which you, as an individual, can reduce gaseous emissions that contribute to acid rain.

9. State the arguments that scientists are making who say that global warming is not due to burning fossil fuels.

10. What are the effects of ground level ozone?

11. Briefly describe the methods by which information is gathered and used to show that the planet is warming up.

multiple choice questions

1. Results of recent climate models suggest that projected greenhouse gas emission patterns may lead to a global warming of 1.5 to 4.5 degrees Celsius during
   a. the next century
   b. the next five years

2. In the last century global surface temperatures have
   a. gone up 1 degree F
   b. gone up 10 degrees F

3. If the air did not contain carbon dioxide, the planet would be
   a. cooler
   b. hotter

4. As a result of global warming, global precipitation would
   a. increase
   b. decrease

5. The energy that causes our atmosphere to heat up comes from
   a. the sun
   b. volcanoes

6. Like carbon dioxide, methane is a greenhouse gas. In the last 150 years the concentration of methane in the atmosphere
   a. stayed the same
   b. increased by 50%
   c. increased by 100%

7. Greenhouse gases in the air around us
   a. are increasing
   b. are decreasing

8. The difference in average temperature between today and the last Ice Age (10,000–12,000 thousand years ago) is
   a. about 2 degrees F
   b. about 9 degrees F

9. Approximately how many tons of carbon dioxide are released into the air by the U.S.?
   a. 6 thousand tons   c. 6 billion tons
   b. 6 million tons   d. 6 trillion tons

10. Global warming occurs when
    a. the total energy leaving the earth is greater than the energy arriving
    b. the heat budget is balanced
    c. infrared light energy is given off by the Earth
    d. the total energy arriving at the Earth is more than the energy leaving

11. CFCs are produced
    a. by humans
    b. by nature
    c. both by humans and nature

12. Methane and nitrous oxide have what impact on the atmosphere?
    a. none at all
    b. allow heat to escape
    c. destroy other greenhouse gases
    d. trap heat

13. Global climate change is due to
    a. the hole in the ozone layer
    b. more UV light coming in
    c. an increase in the concentration of greenhouse gases in the atmosphere due to human activity

14. Critics of global climate change use the following to argue that it is not a serious problem
    a. level of cloud formation
    b. heat trapped by the oceans
    c. higher concentration of CO_2 aids forestation
    d. climate naturally changes or changed in the past

15. Acid rain damages trees by
    a. removing soil nutrients
    b. dissolving roots
    c. causing bark to fall off

16. Most acid rain in the U.S. has a pH of
    a. 8.2   c. 3.4
    b. 7.1   d. 4.2

17. Acid rain can damage your skin.
    a. true
    b. false
    c. only if you swim in an acidic lake

18. Which of the following conditions contribute to acid rain damage?
    a. areas with great soil buffering capacity
    b. areas with poor soil buffering capacity
    c. areas with dense vegetation
    d. areas with sparse vegetation

19. Its formation involves VOCs and NO_x in the presence of sunlight.
    a. ground level ozone
    b. acid rain
    c. carbon monoxide

20. Ozone is typically at the greatest concentrations
    a. at high altitudes (10–15 miles)
    b. upwind of cities (10–15 miles)
    c. downwind of cities (10–15 miles)
    d. well outside of cities (100 miles)

21. Which of the following is not needed to produce ground level ozone?
    a. NO_x
    b. hydrocarbons
    c. SO_2
    d. sunlight

22. Which of the following will reduce vehicular NO_x emissions?
    a. washing the car
    b. rolling down the windows instead of using the AC
    c. keeping the car well tuned
    d. pumping gas at night

23. The stratosphere begins at what altitude?
    a. 1 mile   c. 50 miles
    b. 10 miles   d. 100 miles

24. Ozone is good for the environment at
    a. high altitude (10–15 miles)
    b. low altitude (1–2 miles)
    c. both
    d. neither

25. The bulk of the earth's ozone is in the
    a. troposphere   c. ionosphere
    b. oceans   d. stratosphere

26. The ozone hole
    a. has a constant area throughout the year
    b. has a variable area throughout the year
    c. grew significantly larger this year from last year
    d. shrank significantly this year from last year

27. Ozone is destroyed by
    a. PHDs
    b. VOCs
    c. CFCs

$\mathscr{A}$ppliances

5

chapter

goals

- To understand the operating principles of day-to-day residential appliances

- To be able to read and use Energy Guide labels

- To be able to perform life-cycle analysis of appliances

- To explore ways to save energy and money by understanding good operating practices

Appliance Energy Consumption

Most homes have a variety of appliances with a wide range of operating costs. Typical costs of the operation of basic household appliances are shown in Figure 5.1. These appliances, cooking, and lighting consume 33 percent of the energy at home as shown in Figure 5.2. Water heating (not included in the appliances) is the second largest energy expense after home heating and cooling. It typically accounts for about 14 percent of the utility bill.

Energy Guide Labels

All major home appliances must meet the Appliance Standards Program set by the U.S. Department of Energy (DOE). Manufacturers must use standard test procedures developed by DOE to prove the energy use and efficiency of their products. Test results are printed on yellow Energy Guide labels (Figure 5.3), which manufacturers are required to display on many appliances. This label provides the necessary information to perform a Life Cycle Analysis when comparing different models.

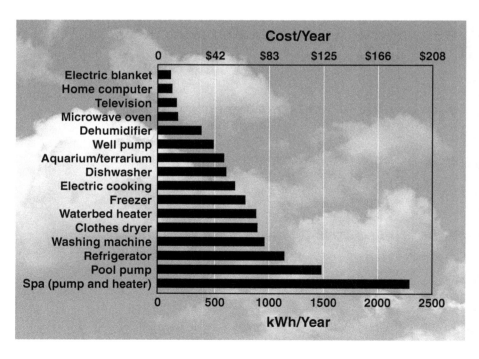

Figure 5.1   *Typical energy costs for various appliances.*

Source: *http://www.eere.energy.gov/consumerinfo/energy_savers/appliance_barchart.html*

The largest portion of a
utility bill for a typical house
is for heating and cooling.

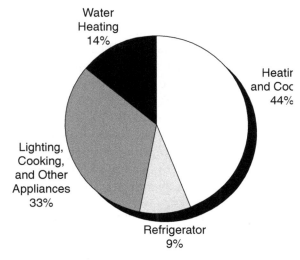

Water
Heating
14%

Heatir
and Coc
44%

Lighting,
Cooking,
and Other
Appliances
33%

Refrigerator
9%

How We Use Energy in Our Homes
(based on national averages)

Figure 5.2

Typical residential energy use.

SOURCE: *www.eere.energy.gov/consumerinfo/energy_savers/energyuse.html*

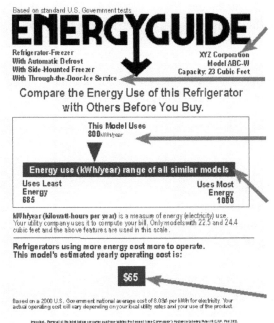

Manufacturer, model number and appliance type.

Information about features, capacity and
size, so you can compare models.

Estimates of the appliance's annual energy
use. The lower the number, the more energy-
efficient the appliance, and the less it costs
to run.

The range of ratings for similar models, from
"uses least energy" to "uses most energy."
This scale shows how a particular model
measures up to the competition.

An estimate of the annual cost to run this model.

Figure 5.3

An example of an Energy Guide label displayed on appliances.

SOURCE: *http://www.ftc.gov/bcp/conline/pubs/homes/applnces.htm*

The Federal Trade Commission's Appliance Labeling Rule requires appliance manufacturers to put these labels on refrigerators, freezers, dishwashers, clothes washers, water heaters, furnaces, boilers, central air conditioners, room air conditioners, heat pumps, and pool heaters. The law requires that the labels specify: The capacity of the particular model—for refrigerators, freezers, dishwashers, clothes washers, and water heaters; the estimated annual energy consumption of the model—for air conditioners, heat pumps, furnaces, boilers, and pool heaters; the energy efficiency rating; and the range of estimated annual energy consumption, or energy efficiency ratings, of comparable appliances. A worksheet on how to use the labels in choosing a cost-effective and environmentally friendly appliance is given below.

PART A—General Information

1. Are the appliances comparable in size and features?   Answer has to be yes
2. What is the price of the more energy-efficient model? $_____
3. What is the price of the lower energy-efficient model? $_____
4. What is the price of electricity in your region? $_____ /kWh
5. How long do you expect to keep the appliance or life
    of the appliance?   _____

PART B—Determining Why You Should Buy an Energy Efficient Model

Calculating the price difference:
1. Price of the more energy-efficient model $_____
2. Price of the lower energy-efficient model $_____
3. Price difference $_____

Determining the annual energy savings:
1. Annual energy consumption of the lower energy-efficient model   _____ kWh
2. Annual energy consumption of the more energy-efficient model   _____ kWh
3. Annual energy savings   _____ kWh

Determining the savings:
1. Annual energy savings   _____ kWh
2. Annual monetary savings on energy (energy savings × price) $_____
3. Energy savings over the lifetime of the appliance
    (Life in years × annual energy savings)   _____ kWh
4. Cost of energy savings over lifetime of the appliance $_____

Determining the payback period:
1. Price difference between the models $_____
2. Annual monetary savings on energy (energy savings × price) $_____
3. Payback period (years to recover the additional investment) $_____
4. Monetary savings on energy over the lifetime $_____
5. Price difference $_____
6. Total monetary benefit for choosing environmentally
    friendly appliance (#4 – #5) $_____

Water Heaters

Heat is continuously flowing from the tank of a water heater and the pipes to the room because the water heater is always at a higher temperature than the surroundings (basement or garage). Thermal energy flows from high temperature to low temperature. Heat is lost whether you use water or not. Like most appliances, water heaters have improved greatly in recent years. Today's models are much more energy efficient, and you will be able to purchase a more efficient water heater that will save you money on energy each month. The average life expectancy of a water heater is 13 years. Therefore, the initial purchase price should not be an important factor in selecting a water heater. Each month you pay for the fuel you use for about 13 years. An energy-efficient model could save hundreds of dollars in the long run in energy costs and may offset the higher initial purchase price. It can be compared to automobile mileage—some cars get 15 miles to a gallon, while other, more efficient, vehicles can go 30 miles or more on a gallon of gas. In the same way, some water heaters use energy more efficiently. One should buy an energy-efficient water heater and spend less money each month to get the same amount of hot water. Table 5.1 shows typical water use for various purposes at home.

Energy Required for Water Heating

Energy required to heat the water is proportional to the temperature difference. If the water comes into the home at 55°F, and needs to be heated to 120°F, then the water needs to be heated by 65°F.

Table 5.1 *Residential Hot Water Use*

USE	GALLONS PER USE
Shower	7–10
Bath (standard tub)	20
Bath (whirlpool tub)	35–50
Clothes washer (hot water wash, warm rinse)	32
Clothes washer (warm wash, cold rinse)	7
Automatic dishwasher	8–10
Food preparation and cleanup	5
Personal (hand-washing, etc.)	2

Heat required is given by

$$Q = m \times C_p \; x \; \Delta T \tag{5.1}$$

where m = mass of water heated, C_p is the heat capacity of water (1 BTU/lb °F) and ΔT = temperature difference.

It is estimated by the United States Department of Energy that a family of four each showering for 5 minutes a day consumes about 700 gal of hot water a week.

Heat energy required to heat 700 gal can be calculated as follows:

$$Heat = \frac{700 \; gal}{week} \times \frac{8.3 \; lb}{gal} \times 1\frac{BTU}{lb \; °F} \times (120°F - 55°F) = 377,650\frac{BTU}{Week}$$

The heat requirement is 377,650 BTU/week × 52 weeks/year = 19,637,800 BTU/year or it is also equal to 5,755 kWh.

Assuming that natural gas costs $10/MMBTU or 0.092 per kWh, the gas costs would be $196.37 and the electric costs would be $529.46. Clearly, electric heat is more expensive than natural gas.

Energy costs increase with water temperature. Dishwashers require the hottest water of all household uses, typically 135°F to 140°F. However, these devices are usually equipped with booster heaters to increase the incoming water temperature by 15°F to 20°F. Setting the water heater between 120°F and 125°F and turning the dishwasher's booster on should provide sufficiently hot water while reducing the chances for scalding.

Types of Water Heaters

There are several types of water heaters available in the market: storage or tank type, on demand, heat pump, tankless coil, indirect, and solar water heaters. However, most water heaters use a storage tank type. (See Illustration 5.1)

Storage or Tank-Type Water Heaters

Storage or tank-type water heaters are relatively simple devices and by far the most common type of residential water heater used in the United States. They range in size from 20 to 80 gallons and can be fueled by electricity, natural gas, propane, or oil. When you turn on a hot water faucet or use hot water in a dishwasher or clothes washer, water pipes draw hot water from the tank. As can be seen in Figure 5.4, to replace that hot water, cold water enters the bottom of the tank, ensuring that the tank is always full. Depending on the type of fuel used, either electrical heating elements or a natural gas burner is used to heat the water. Electric water heaters are generally less expensive to install (purchase price) than gas-fired types because they don't require gas lines and vents to let the combustion products out of the house. Natural gas costs about $7 to $12 per million BTUs whereas electrical energy is $20 to $25 per million BTUs, making electric water heaters more expensive to operate.

Illustration 5.1

Estimate the % energy savings of an electric water heater that heats 100 gallons of water per day when the temperature is set back at 110° instead of 120°F. The basement is heated and is at 65°F. The life of the water heater is expected to be about 10 years. Use an appropriate cost for electricity and compare the operating expenses with the approximate initial cost of the water heater.

$$\text{Heat required (BTU)} = m \times C_p \times (\text{Temperature difference})$$

where C_p is the heat capacity of water (1 BTU/lb/F) and m is the mass of the water. (Assume 1 gal has 8.3 lb of water and that 3,412 BTUs = 1 kWh.)

Solution: **Energy required for heating the water to 120°F**

$$= m \times C_p \times \Delta T$$

$$= \underbrace{\frac{100\ gal}{day} \times \frac{8.3\ lb}{gal}}_{m} \times \underbrace{\frac{1\ BTU}{lb\ °F}}_{C_p} \times \underbrace{(120 - 65)°F}_{\Delta T} = 45,650\ BTUs\ per\ day$$

In a year the energy required is $\dfrac{45,650\ BTU}{day} \times \dfrac{365\ days}{year} = 16,662,250\ BTUs$

In a 10-year period, the energy required is 166,622,500 BTUs, which is equal to 48,834 kWh (166,622,500 BTUs/3412 BTUs/kWh).

Operating cost over its lifetime $= \dfrac{48,834\ kWh}{} \times \dfrac{\$0.09}{kWh} = \$4,395$

Energy required for heating the water to 110°F

$$= m \times C_p \times \Delta T$$

$$= \underbrace{\frac{100\ gal}{day} \times \frac{8.3\ lb}{gal}}_{m} \times \underbrace{\frac{1\ BTU}{lb\ °F}}_{C_p} \times \underbrace{(110 - 65)°F}_{\Delta T} = 37,350\ BTUs\ per\ day$$

In a year the energy required is $\dfrac{37,350\ BTU}{day} \times \dfrac{365\ days}{year} = 13,632,750\ BTUs$

In a 10-year period, the energy required is 136,327,500 BTUs, which is equal to 39,995 kWh (136,327,750 BTUs/3412 BTUs/kWh).

% Energy Savings $= \dfrac{48,834 - 39,995}{48,834} \times 100 = 18.1\%$. It will save \$799 over 10 yrs.

Operating cost over its lifetime $= \dfrac{39,995\ kWh}{} \times \dfrac{\$0.09}{kWh} = \$3,596$

These operating costs are approximately 10 times higher than the initial price we pay.

Electric Hot Water Heater

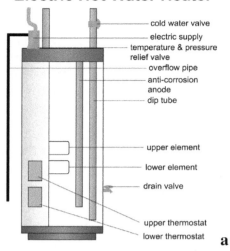

cold water valve
electric supply
temperature & pressure relief valve
overflow pipe
anti-corrosion anode
dip tube

upper element

lower element

drain valve

upper thermostat
lower thermostat

a

Gas Hot Water Heater

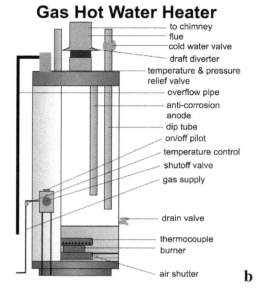

to chimney
flue
cold water valve
draft diverter
temperature & pressure relief valve
overflow pipe
anti-corrosion anode
dip tube
on/off pilot
temperature control
shutoff valve
gas supply

drain valve

thermocouple
burner

air shutter

b

Figure 5.4   *Internals of a storage or tank-type water heater: (a) electric (b) gas.*

Storage tank-type water heaters raise and maintain the water temperature to the temperature setting on the tank (usually between 120°–140°F). Because the water is constantly heated and kept ready for use in the tank, heat energy can be lost even when no faucet is on. This is called standby heat loss. These standby losses represent 10 to 20 percent of a household's annual water heating costs. Newer, more energy-efficient storage models can significantly reduce the amount of standby heat loss, making them much less expensive to operate.

Demand Water Heaters

It is possible to completely eliminate standby heat losses from the tank and reduce energy consumption by 20 to 30 percent with on-demand (or instantaneous) water heaters. These water heaters do not have storage tanks. Demand water heaters are available in propane (LP), natural gas, or electric models. Cold water travels through a pipe into the unit, and either a gas burner or an electric element heats the water only when needed (Figure 5.5). With these systems, you never run out of hot water. However, the flow rate is limited by the outlet temperature.

Typically, demand heaters provide hot water at a rate of 2 to 4 gallons per minute. This flow rate might meet the requirements of a household's hot water needs, but not more than one location at the same time (e.g., showering and doing laundry simultaneously). To meet hot water demand when multiple faucets are being used, demand heaters can be installed in parallel sequence. Although gas-fired demand heaters tend to have higher flow rates than electric ones, they can waste energy even when no water is being heated if their pilot

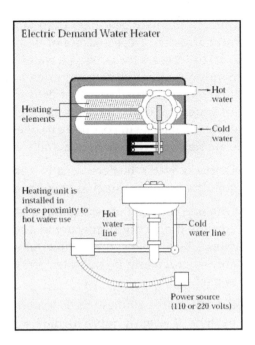

Electric Demand Water Heater

Heating elements

Hot water

Cold water

Heating unit is installed in close proximity to hot water use

Hot water line

Cold water line

Power source (110 or 220 volts)

Figure 5.5

A schematic diagram of a tankless or demand water heater.
SOURCE: *http://www.eere.energy.gov/consumerinfo/pdfs/watheath.pdf*

lights stay on. However, the amount of energy consumed by a pilot light is quite small. Demand water heaters cost more than conventional storage tank-type units. Small point-of-use heaters that deliver 1 to 2 gpm sell for about $200. Larger gas-fired demand units that deliver 3 to 5 gpm cost $550–$1,000.

The appeal of demand water heaters is not only the elimination of the tank standby losses and the resulting lower operating costs, but also the fact that the heater delivers hot water continuously. Gas models with a standing (constantly burning) pilot light, however, offset some of the savings achieved by the elimination of tank standby losses with the energy consumed by the pilot light.

Most demand models have a life expectancy of more than 20 years. In contrast, storage tank water heaters last 10 to 15 years. Most demand models have easily replaceable parts that can extend their life by many years more.

Advantages and Disadvantages

Demand water heaters are compact in size and virtually eliminate standby losses—energy wasted when hot water cools down in long pipes or while it's sitting in the storage tank.

By providing warm water immediately where it's used, demand water heaters waste less water. People don't need to let the water run as they wait for warm water to reach a remote faucet. A demand water heater can provide unlimited hot water as long as it is operating within its capacity.

Equipment life may be longer than tank-type heaters because they are less subject to corrosion. Expected life of demand water heaters is 20 years, compared to 10 to 15 years for tank-type water heaters.

Demand water heaters range in price from $200 for a small under-sink unit up to $1,000 for a gas-fired unit that delivers 5 gallons per minute. Typically, the more hot water the unit produces, the higher the cost. In most cases, electric demand water heaters will cost more to operate than gas demand water heaters.

Here are some drawbacks to demand water heating:

- Demand water heaters usually cannot supply enough hot water for simultaneous uses such as showers and laundry.

- Unless your demand system has a feature called modulating temperature control, it may not heat water to a constant temperature at different flow rates. That means that water temperatures can fluctuate uncomfortably—particularly if the water pressure varies wildly in your own water system.

- Electric units will draw more instantaneous power than tank-type water heaters. If electric rates include a demand charge, operation may be expensive.

- Electric demand water heaters require a relatively high electric power draw because water must be heated quickly to the desired temperature. Make sure your wiring is up to the demand.

- Demand gas water heaters require a direct vent or conventional flue. If a gas-powered unit has a pilot light, it can waste a lot of energy.

Solar Water Heaters

An estimated one million residential and 200,000 commercial solar water-heating systems have been installed in the United States. Although there are a large number of different types of solar water-heating systems, the basic technology is very simple. Sunlight strikes and heats an "absorber" surface within a "solar collector" or an actual storage tank. These roof-mounted solar heaters (shown in Figure 5.6) supply about 80 percent of the hot water for the home. Either a heat-transfer fluid or the actual potable water to be used flows through tubes attached to the absorber and picks up the heat from it. (Systems with a separate heat-transfer-fluid loop include a heat exchanger that then heats the potable water.) The heated water is stored in a separate preheat tank or a conventional water heater tank until needed. If additional heat is needed, it is provided by electricity or fossil-fuel energy by the conventional water-heating system. By reducing the amount of heat that must be provided by conventional water heating, solar water-heating systems directly substitute renewable energy for conventional energy, reducing the use of electricity or fossil fuels by as much as 80 percent.

Figure 5.6   *A roof-mounted solar water heater.*
SOURCE: *http://www.eere.energy.gov/buildings/info/components/waterheating/solarhot.html*

Today's solar water-heating systems are proven reliable when correctly matched to climate and load. The current market consists of a relatively small number of manufacturers and installers that provide reliable equipment and quality system design. A quality assurance and performance-rating program for solar water-heating systems, instituted by a voluntary association of the solar industry and various consumer groups, makes it easier to select reliable equipment with confidence. Building owners should investigate installing solar hot water-heating systems to reduce energy use. Before sizing a solar system, water-use reduction strategies should be put into practice.

There are five types of solar hot-water systems:

1. **Thermosiphon Systems.** These systems heat water or an antifreeze fluid, such as glycol. The fluid rises by natural convection from collectors to the storage tank, which is placed at a higher level. No pumps are required. In thermosiphon systems fluid movement, and therefore heat transfer, increases with temperature, so these systems are most efficient in areas with high levels of solar radiation.

2. **Direct-Circulation Systems.** These systems pump water from storage to collectors during sunny hours. Freeze protection is obtained by recirculating hot water from the storage tank or by flushing the collectors (drain-down). Since the recirculation system increases energy use while flushing reduces the hours of operation, direct-circulation systems are used only in areas where freezing temperatures are infrequent.

3. **Drain-Down Systems.** These systems are generally indirect water-heating systems. Treated or untreated water is circulated through a closed loop, and heat is transferred to potable water through a heat exchanger. When no solar heat is available, the collector fluid is drained by gravity to avoid freezing and convection loops in which cool collector water reduces the temperature of the stored water.

4. **Indirect Water-Heating Systems.** In these systems, freeze-protected fluid is circulated through a closed loop and its heat is transferred to potable water through a heat exchanger with 80 to 90 percent efficiency. The most commonly used fluids for freeze protection are water-ethylene glycol solutions and water-propylene glycol solutions.

5. **Air Systems.** In this indirect system the collectors heat the air, which is moved by a fan through an air-to-water heat exchanger. The water is then used for domestic or service needs. The efficiency of the heat exchanger is in the 50 percent range. Direct-circulation, thermosiphon, or pump-activated systems, require higher maintenance in freezing climates. For most of the United States, indirect air and water systems are the most appropriate. Air solar systems, while not as efficient as water systems, should be considered if maintenance is a primary concern since they do not leak or burst.

Heat Pump Water Heaters

Heat pump water heaters can provide up to 60 percent energy savings over conventional water heaters. Figure 5.7 shows a schematic of a heat pump water heater principle.

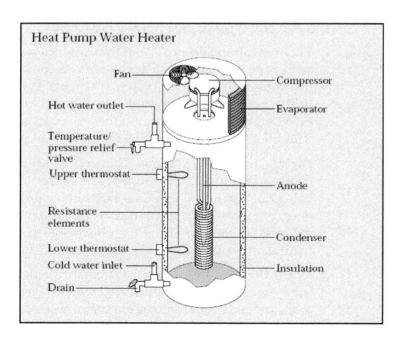

Figure 5.7

A schematic of a heat pump water heater.

SOURCE: *http://www.eere.energy.gov/consumerinfo/pdfs/watheath.pdf*

Heat pumps are a well-established technology for space heating. The same principal of transferring heat is at work in heat pump water heaters (HPWHs) except that they extract heat from air (indoor, exhaust, or outdoor air) and deliver it to water. Some models come as a complete package, including tank and back-up resistance heating elements, while others work as an adjunct to a conventional water heater. Because it extracts heat from air, the HPWH delivers about twice the heat for the same electricity cost as a conventional electric resistance water heater.

The simplest HPWH is the ambient air-source unit, which removes heat from surrounding air, providing the additional benefit of space cooling. Exhaust air units extract heat from a continuously exhausted air stream and work better in heating-dominated climates because they do not cool ambient air. Some units can even be converted between the two modes of operation for optimum operation in either summer or winter. In mild climates you can locate units in unheated but protected spaces such as garages, essentially using outdoor air as a heat source.

A variation of the stand-alone HPWH is the desuperheater feature available on some central air conditioners. It provides economical supplemental water heating as a byproduct of air conditioning. Desuperheater water heating can be part of an integrated package with a heat pump or air conditioner system. In most such systems, the heat pump water heating only occurs during normal demand for space conditioning, with resistance electric coils providing water heating the rest of the time. During the cooling season, the desuperheater actually improves the efficiency of the air conditioning system while heating water at no direct cost. In an average climate, a desuperheater might meet 20 to 40 percent of annual water heating demand.

Most of the heat delivered to the water comes from the evaporator of the unit, not through the electrical input to the machine. Consequently, the efficiency of the HPWH is much higher than for direct-fired gas or electric storage water heaters. The installed cost of commercial HPWH systems is typically several times that of gas or electric water heaters, yet the low operating costs can often offset the higher total installed cost, making the HPWH the economic choice for water heating. The HPWH becomes increasingly attractive in building applications where energy costs are high and where there is a steady demand for hot water. This attractiveness is less a function of building type than it is of water demand and utility cost.

Energy Efficiency of Water Heaters

The federal efficiency standards for water heaters took effect in 1990, assuring consumers that all new water heaters meet certain minimum-efficiency levels. New standards, which took effect in January 2004, will increase the minimum efficiency levels of these products. Water heater efficiency is reported in terms of the energy factor (EF). EF is an efficiency ratio of the energy supplied in heated water divided by the energy input to the water heater. The higher the EF, the more efficient the water heater.

Electric resistance water heaters have EFs ranging from 0.7 and 0.95; gas water heaters range from 0.5 and 0.6, with some high-efficiency models ranging around 0.8; oil water heaters range from 0.7 and 0.85; and heat-pump water heaters range from 1.5 to 2.0. Everything else being equal, select a water heater with the highest energy factor (EF). Also look for a water heater with at least one-and-a-half inches of tank insulation and a heat trap. There is little difference between the most efficient electric resistance storage water heaters and the minimum-efficiency standard that took effect in January 2004. If you need to rely on electricity to heat your water, keep your eye out for the further development of heat-pump water heaters. This technology uses one-third to one-half as much electricity as a conventional electric resistance water heater. The efficiency of water heaters is indicated by their energy factor (EF), which is based on recovery efficiency, standby losses, and cycling losses. The higher the EF, the more efficient the water heater. Table 5-2 lists the recommended and best available energy factors for electrical water heaters.

Capacity of a water heater is an important consideration. The water heater should provide enough hot water at the busiest time of the day. For example, a household of two adults may never use more than 30 gallons of hot water in an hour, but a family of six may use as much as 70 gallons in an hour. The ability of a water heater to meet peak demands for hot water is indicated by its "first hour rating." This rating accounts for the effects of tank size and the speed by which cold water is heated. Water heaters must be sized properly. Oversized water heaters not only cost more but increase energy use due to excessive cycling and higher standby losses.

Let's use the Energy Guide to perform a life-cycle analysis and choose a water heater. Different models of water heaters with the same capacity can vary dramatically in the amount of electricity they use. For a 40-gallon water heater, for example, the annual electricity consumption for models A and B was 4,622 kilowatt-hours and 4,989 kilowatt-hours a year, respectively.

Table 5.2   *Energy Efficiency Recommendations*

STORAGE TYPE	RECOMMENDED		BEST AVAILABLE	
	ENERGY FACTOR	ANNUAL ENERGY USE (kWh)	ENERGY FACTOR	ANNUAL ENERGY USE (kWh)
Less than 60 gallons	0.93	4,721	0.95	4,622
60 gallons or more	0.91	4,825	0.92	4,773

How to Use Energy Guide Labels—Comparing Two Water Heaters

PART A—General Information

1.	Are the appliances comparable in size and features?	Yes
2.	What is the price of the more energy-efficient model?	$234
3.	What is the price of the lower energy-efficient model?	$189
4.	What is the price of electricity in your region?	$0.092/kWh
5.	How long do you expect to keep the appliance or life of the appliance?	13 years

PART B—Determining Why You Should Buy an Energy Efficient Model

Calculating the price difference:

6.	Price difference	$45

Determining the annual energy savings:

7.	Annual energy consumption of the lower energy-efficient model	4,989 kWh
8.	Annual energy consumption of the more energy-efficient model	4,622 kWh
9.	Annual energy savings	367 kWh

Determining the savings:

10.	Annual monetary savings on energy (energy savings × price)	$33.76
11.	Energy savings over the lifetime of the appliance (Life in years × annual energy savings)	4,771 kWh
12.	Cost of energy savings over lifetime of the appliance	$438.96

Determining the payback period:

13.	Price difference between the models	$45
14.	Annual monetary savings on energy (energy savings × price)	$33.76
15.	Payback period (years to recover the additional investment)	1.33 years
16.	Monetary savings on energy over the lifetime	$438.96
17.	Price difference	$45
18.	Total monetary benefit for choosing environmentally friendly appliance (#16 – #17)	$434.96

Obviously it pays to buy an energy-efficient water heater and save $434.96. It also helps the environment by not using 4,771 kWh of electrical energy and thereby not emitting 9,652 lb of CO_2, 21 lb of NO_x, 75 lb of SO_2 and 1 lb of CO and particulate matter each into the environment. See the individual's power!

Water Heaters: Your "Power" in the Environmental Protection

☞ Do as much cleaning as possible with cold water to save the energy used to heat water.

☞ Check your faucets for leaks. They waste both water and energy!

- Conserve hot water by installing water-saving showerheads. A new showerhead can save as much as $10 a year in water and energy.

- Once your water is hot, insulate to help keep it that way. Wrapping exposed hot water pipes with insulation will minimize heat loss. So will installing an R-12 insulation blanket around your water heater, unless the manufacturer does not recommend it.

- Reduce your water heater's temperature to 120°F. That will produce plenty of hot water and still save energy. For homes with a dishwasher, a setting of 140°F is required to clean properly, but most of the new dishwashers have a built-in water temperature booster.

- Many new water heaters have a "vacation" setting you can use to save energy if you're away for more than a few days. Turn the thermostat down or off when you're gone for more than three days.

Refrigerators

Refrigerators are heat movers, which move heat from a low temperature (inside the refrigerator) to a high temperature (outside the refrigerator into the kitchen). Heat movers do not produce any heat but just move it from one location to another.

How a Refrigerator Works

The principle of operation of a refrigerator is similar to that of an air conditioner. They both do not move air but move the heat energy from inside to outside. There are four basic components in a refrigerator (Figure 5.8), and their functions are as follows: A liquid refrigerant at high pressure flows through an **expansion valve.** The high pressure refrigerant expands in the expansion valve to low pressures. During the expansion, the temperature of the refrigerant decreases, thereby producing cold temperatures. This cold refrigerant flows through an **evaporator** or **heat exchanging pipes** in the refrigerator, absorbing the heat from the items inside. Remember that heat always flows from high temperature to low temperature. In the evaporator, the liquid refrigerant evaporates by absorbing heat from the food and hence the name "evaporator" coils. The food gets cold as a result. The evaporated refrigerant is kind of warm and goes through a **compressor**—a device that pressurizes the warm refrigerant and makes it hot (hotter than the kitchen temperature). This hot refrigerant goes into the **condenser** or **second heat exchanger coils**, located at the back of the refrigerator, where it gives off the heat to the air in the kitchen.

Refrigerator

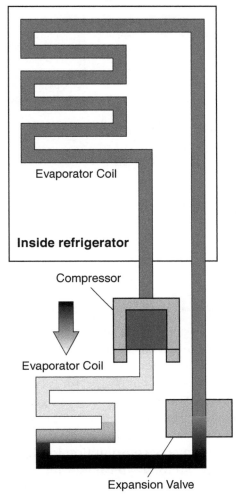

Evaporator Coil

Inside refrigerator

Compressor

Evaporator Coil

Expansion Valve

Main components of a refrigerator.

Types and Features

There are four types of refrigerators: top-freezer (or top-mount), bottom-freezer (or bottom-mount), side-by-side, and built-in (shown in Figure 5.9). Refrigerators also come in four size categories: small (7 to 9.9 cubic feet), medium (10 to 13.9 cubic feet), large (14 to 19.9 cubic feet), and extra large (20 to 29 cubic feet). Can you think which type is more efficient and why?

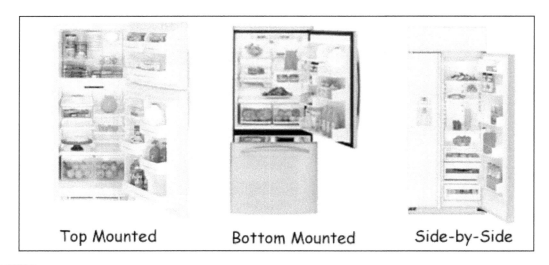

Top Mounted   Bottom Mounted   Side-by-Side

Types of refrigerators.

Energy Efficiency of a Refrigerator

Refrigerators come with an Energy Guide label that tells you in kilowatt-hours (kWh) how much electricity a particular model uses in a year. The smaller the number, the less energy the refrigerator uses and the less it will cost you to operate. The efficiency of a refrigerator is based on the energy consumed per year for a given size. The efficiency of a refrigerator is expressed in "volume cooled per unit of electric energy per day." Volume is measured in cubic feet, and electrical energy is measured in kilowatt hours. The DOE standards set maximum allowable annual energy consumption for different sizes and classes of refrigerators. The energy bill for a typical new refrigerator with automatic defrost and top-mounted freezer will be about $55/year, whereas a typical model sold in 1973 used to cost nearly $160/year (almost three times the energy consumption). Most of the energy used by a refrigerator is used to pump heat out of the cabinet. A small amount is used to keep the cabinet from sweating, to defrost the refrigerator, and to illuminate the interior.

The energy efficiency of refrigerators and freezers has improved dramatically over the past three decades. Federal efficiency standards first took effect in 1993, requiring new refrigerators and freezers to be more efficient than ever before. A new set of stricter standards took effect July 1, 2001. Full-sized refrigerators that exceed the federal standard by 15 percent or more (and full-sized freezers that exceed it by 10 percent) qualify for the ENERGY STAR label. Compact refrigerators and freezers must exceed the standard by 20 percent to qualify for ENERGY STAR.

ENERGY STAR qualified refrigerators provide energy savings without sacrificing the features you want. ENERGY STAR–qualified models use high-efficiency compressors, improved insulation, and more precise temperature and defrost mechanisms to improve energy efficiency. These models also use at least 15 percent less energy than required by current federal standards and 40 percent less energy than the conventional models sold in 2001. Many ENERGY STAR qualified models include automatic ice-makers and through-the-door ice dispensers. Qualified models are also available with top, bottom, and side-by-side freezers.

Refrigerators with the freezer on either the bottom or top are the most efficient. Bottom freezer models use approximately 16 percent less energy than side-by-side models, and top freezer models use about 13 percent less than side-by-side.

Through-the-door icemakers and water dispensers are convenient and reduce the need to open the door, which helps maintain a more constant temperature; however, these convenient items will increase your refrigerator's energy use by 14 to 20 percent. Mini-doors give you easy access to items most often used. The main door is opened less often, which saves energy.

Too large a refrigerator may waste space and energy. One that's too small can mean extra trips to the grocery store. Your best bet is to decide which size fits your needs and then compare the Energy Guide label on each so you can purchase the most energy efficient make and model.

A manual defrost refrigerator uses half the energy of an automatic defrost model but must be defrosted regularly to stay energy efficient.

Refrigerators with anti-sweat heaters consume 5 percent to 10 percent more energy. Look for models with an "energy saver" switch that lets you turn down—or off—the heating coils (which prevent condensation).

Technology Improvements

The improvement in energy efficiency over the past three decades is due to the addition of vacuum insulation panels around the freezer section to reduce heat transfer and the addition of polyurethane foam to the doors to double insulation thickness. Replacement of AC motors with more efficient DC motors and replacement of automatic defrost control with an adaptive defrost that operates only when needed have also contributed to the increase in efficiency.

Refrigerators: Your "Power" in the Environmental Protection

☞ Keep your refrigerator or freezer at the following temperatures: 37–40°F for the fresh food compartment of the refrigerator, 0–5°F for the freezer section. Use a thermometer to check inside temperatures.

- Regularly defrost manual-defrost refrigerators and freezers; don't allow frost to build up more than ¼ inch.

- Make sure your refrigerator and freezer door seals are airtight. Check the seal on door gaskets periodically by closing the door on a dollar bill. If it pulls out easily, you may need a new gasket.

- Keep the doors closed as much as possible and make sure they are closed tightly.

- To ensure proper cooling of its contents, don't crowd food items. Too many dishes obstruct air circulation.

- Cover liquids and wrap foods stored in the refrigerator. Uncovered foods release moisture and make the compressor work harder.

- Replace paper wrappings on food items with aluminum foil or plastic wrap. Paper is an insulator.

- Experiment with the "energy saver" switch in your refrigerator—it allows you to adjust the heating coil under the "skin" of the refrigerator (the purpose of the heating coils is to prevent condensation on your refrigerator).

- Placement of the refrigerator is very important. Direct sunlight and close contact with hot appliances will make the compressor work harder. More importantly, heat from the compressor and the condensing coil must be able to escape freely, or it will cause the same problem. Don't suffocate the refrigerator by enclosing it tightly in cabinets or against the wall. The proper breathing space will vary depending on the location of the coils and compressor on each model—something important to know before the cabinets are redesigned.

- Regularly brush off or vacuum the refrigerator coils on the back or bottom of the unit.

- Because most refrigerators reject heat from the bottom and/or back, they need adequate clearance to allow sufficient airflow. While no specific studies have been done to calculate the optimum clearance space, one general rule of thumb is to double the space recommended by manufacturers for refrigerator installation. Another rule-of-thumb is to allow two inches of air flow around the refrigerator.

- Don't keep that old, inefficient fridge running day and night in the garage for those few occasions when you need extra refreshments. A 15-year-old refrigerator could cost $100–$150 per year in operating expenses.

Clothes Washers and Dryers

Clothes washers and dryers account for 10 percent of the residential energy consumption. Most of the energy consumed by clothes washers is for hot water used for washing. It is estimated that 85 percent to 90 percent of the energy is used for heating the water. Relatively 10 percent to 15 percent of the energy is used by the clothes washer itself to operate the motor and controls. A typical household does nearly 400 loads of laundry a year, and each load in a conventional washer uses 40 gallons of water. Therefore, any reduction in energy consumption for clothes washing application would involve reduction in hot water use.

The basic principle for cleaning clothes has remained unchanged—wet the garment, agitate it to loosen the dirt from the cloth fibers and then use more water to rinse the dirt off. What has changed over the millennia is the method of agitation. Pounding garments with stones was common for several thousand years, and along the way someone also figured out that using heated water got out a lot more dirt.

Clothes washers come in two types: Horizontal axis or front loading and vertical axis or top loading (shown in Figure 5.10).

Most clothes washers produced for the U.S. consumer are top loading washers with a central agitator. While there are variations, most top loading washers suspend the clothes in a tub of water for washing and rinsing. In contrast, the front loading washer tumbles the wash load repeatedly through a small pool of water at the bottom of the tub to produce the needed agitation. This tends to reduce the need for both hot and cold water. The front loading washer, popular in Europe, has a somewhat limited market share in the United States at present.

a

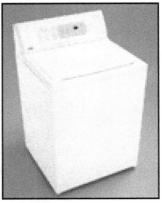
b

Figure 5.10 *Picture of two types of clothes washers: (a) Horizontal axis (front loading) and (b) Vertical axis or top loading.*

Source: *http://www.energystar.gov/index.cfm?c=media_kit.nr_news_photos2*

But its share is growing. Estimates have shown that a large quantity of energy and water could be saved through the replacement of conventional top loading washers with the front loading design. Front loading machines repeatedly lift and tumble clothes, instead of moving clothes around a central axis. Front loading washers also use sensor technology to closely control the incoming water temperature. To reduce water consumption, they spray clothes with repeated high-pressure rinses to remove soap residues rather than soaking them in a full tub of rinse water.

A study conducted by Oak Ridge National Laboratory (ORNL) in 1998 for the U.S. Department of Energy found that, on average, the front loading washer used 62.2 percent of the water used by a top loading washer, which yielded a total water savings of 37.8 percent. Moreover, the average front loading washer consumed 42.4 percent of the energy used by a typical top loading washer in the study, resulting in an energy savings of 57.6 percent.

Auto Temperature

The machine will mix hot and cold water to a preset "warm" and "cold" so that the water is warm enough for the detergent to dissolve and optimally perform. During the winter in many parts of the country, cold tap water can be too cold to wash your clothes well.

Water Level Settings

Some of the very high-end machines sense the amount of clothing and automatically adjust the water level, but all except the most basic machines offer at least four settings.

Capacity

If you have a large household or athletes who produce an astounding amount of laundry each week, a larger capacity machine is a must. The front loaders generally hold more because they don't have the agitator. A definite minus for the front loaders, however, is the actual loading because you have to bend over to put in the clothes. To minimize this fact, the machines and their matching dryers are often displayed in stores on a raised platform.

Energy Efficiency and Water Usage

Since the front loaders use less water, they use less energy. The most efficient front loaders use less than half the amount of water used in the average top loaders. Simply washing with cold water, however, will conserve energy and is also recommended for colored and many delicate fabrics.

Washing machines with the ENERGY STAR designation are 50 percent more energy efficient than the current minimal allowable standard. Many new energy-efficient, water-conserving clothes washers have been introduced over the past few years. These resource-efficient washers are available in a variety of sizes and configurations offering consumers a wide range of front-loading and top-loading styles in many different price ranges.

Energy Factor is a metric that was previously used to compare relative efficiencies of clothes washers. The higher the Energy Factor is, the more efficient the clothes washer is. For clothes washers, Energy Factor is calculated using the following formula:

$$Energy\ Factor = \frac{392 \times Volume\ (ft^3)}{Annual\ energy\ use\ (kwh)}$$

It is based on an average usage of 392 loads of laundry per year. Your actual energy consumption will depend on your number of loads.

Water Factor is the number of gallons per cycle per cubic foot that the clothes washer uses. The lower the water factor, the more efficient the washer is. So, if a clothes washer uses 30 gallons per cycle and has a tub volume of 3.0 cubic feet, then the water factor is 10.0. Most full-sized ENERGY STAR qualified washers use 18–25 gallons of water per load, compared to the 40 gallons used by a standard machine. The lower the water factor, the less water the machine uses.

The Energy Factor for a washer does not indicate the real energy efficiency because of the tub size and other factors. Therefore, the Energy Factor is modified to include the tub size and drying characteristics.

Modified Energy Factor (MEF) is a new equation that replaced Energy Factor as a way to compare the relative efficiency of different units' clothes washers. The equation is shown below:

$$MEF = \frac{C}{M + E + D}$$

where C = capacity of the clothes container
       M = machine electrical consumption
       E = hot water energy consumption
       D = energy required for removal of the moisture in the wash load

The higher the value, the more efficient the clothes washer is. On January 1, 2007 the ENERGY STAR criteria for clothes washers changed. To qualify for ENERGY STAR, a clothes washer must have a minimum Modified Energy Factor (MEF) of 1.72 and also a maximum Water Factor of 8.0. Before January 1, 2007, the minimum MEF was 1.42, and there was not a Water Factor requirement

Clothes Washers: Your "Power" in the Environmental Protection

☛ Wash full loads—Clothes washers are most efficient when operated with full loads.

- Wash clothes in cold water. A substantial amount of the energy used for clothes washing is for heating the water. Selecting cold water wash cycles will save energy; appropriate load-size settings will save both water and energy.

- ENERGY STAR-qualified clothes washers use less water than standard models while cleaning just as well.

- In areas with limited water supplies, consider buying products with low water factors (WF).

Dryers

A clothes dryer dries wet clothes in a rotating drum through which hot air is circulated. The hot air removes the residual moisture from the clothes. The humid air from the dryer is vented out of the house. The drum is rotated with the help of a motor at relatively slow speeds to create a tumbling effect. They can be of two types: electric and gas. In an electric dryer, electrical energy is used both for the motor to rotate the drum and for heating the air. In a gas dryer, the motor requires electrical energy but the air is heated by natural gas.

Energy Efficiency of Dryers

Dryers work by heating and aerating clothes. The efficiency of a clothes dryer is measured by a term called the Energy Factor. It is similar to the miles per gallon for a car, but in this case the measure is pounds of clothing per kilowatt-hour of electricity. The minimum rating for a standard capacity electric dryer is 3.01. For gas dryers the minimum energy factor is 2.67. The rating for gas dryers is provided in kilowatt-hours though the primary source of fuel is natural gas.

Unlike most other types of appliances, energy consumption does not vary significantly among comparable models of clothes dryers. Clothes dryers are not required to display Energy Guide labels.

Clothes Dryers: Your "Power" in the Environmental Protection

- Locate your dryer in a heated space. Putting it in a cold or damp basement will make the dryer work harder and less efficiently.

- Make sure your dryer is vented properly. If you vent the exhaust outside, use the straightest and shortest metal duct available. Flexible vinyl duct isn't recommended because it restricts the airflow, can be crushed, and may not withstand high temperatures from the dryer.

- Check the outside dryer exhaust vent periodically. If it doesn't close tightly, replace it with one that does to keep the outside air from leaking in. This will reduce heating and cooling bills.

- Clean the lint filter in the dryer after every load to improve air circulation. Regularly clean the lint from vent hoods.

- Dry only full loads, as small loads are less economical; but do not overload the dryer.

- When drying, separate your clothes and dry similar types of clothes together. Lightweight synthetics, for example, dry much more quickly than bath towels and natural fiber clothes.

- Dry two or more loads in a row, taking advantage of the dryer's retained heat.

- Use the cool-down cycle (permanent press cycle) to allow the clothes to finish drying with the residual heat in the dryer.

- In good weather, hang clothes outside to dry. This is the ultimate energy saver for clothes drying.

Dishwashers

A dishwasher typically uses the equivalent of 700–850 kilowatt-hours of electricity annually, or nearly as much energy as a clothes dryer or freezer. About 80 percent of this energy is used not to run the machine but to heat the water for washing the dishes.

How a Dishwasher Works

A dishwasher is essentially an insulated watertight box. The dirty dishes are systematically arranged in the dishwasher. Hot water is sprayed onto the dishes as jets, and repeated jets of water emanating from a spray arm, clean the dishes (Figure 5.11). Some models have two spray arms, one at the bottom of the dishwasher (lower spray arm) and one at the top (upper spray arm). The dirty water passes through a filter and re-circulates until the dishes are finished. Fresh water is then sprayed in the rinse cycle to remove the soapy water, and then the dishes are dried with either electric heat or simply air. Older dishwashers use about 8–14 gallons of water for a complete wash cycle. Dishwashers built in the past 10 years have been using 7–10 gallons per cycle. A dishwasher is the only device at home that requires about 140°F. The units built recently have supplemental heaters in the dishwashers to bump up the temperature so that the main water heater temperature can be set at 120°F or less. Remember that each 10°F reduction in water heater temperature lowers the water heater energy cost by 3 percent to 5 percent.

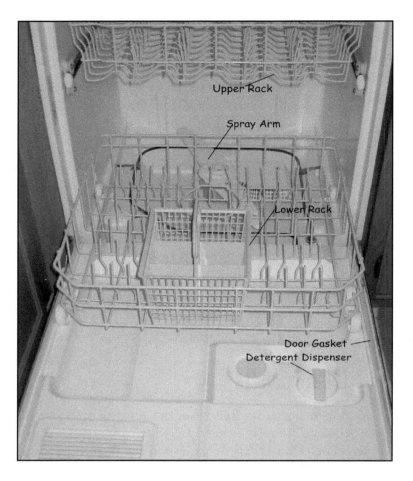

Figure 5.11

A schematic of the internals of a dishwasher.

Dishwashers can be built-in or portable. Built-ins are mounted under a kitchen countertop usually next to a sink. Portables are on wheels with finished tops and sides. Most models can be converted into under-counter mounting. However, because of the additional connection hardware and finished sides, portables usually cost more than similar built-in models. Some of the additional features that are offered are:

Interior layout—Layout includes the configuration of sliding racks, baskets, and trays. Does the washing arm reduce the amount of loadable space?

Water heating—Most homes have water heaters set to 110°F. However, to clean well a dishwasher should use water at 140°F. Many budget units now offer this feature.

Number of cycles—Cycles include light cycles, normal, heavy or pans, and rinse and hold to remove food if dishes will sit in the washer a while before the wash cycle is run.

Water-saving cycles—If you live in an area where fresh water is scarce, you'll want to consider this feature.

Sound insulation—The sound level will vary from one model to another. Consider how important a quiet wash cycle is before you purchase.

Built-in food disposers—Disposers will grind up food in a manner similar to in-sink units, allowing the user to spend less time cleaning dishes before they go in.

Controls—Entry-level machines feature knob and dial controls. On mid- to upper-end models you'll find push-button switches hidden behind smooth one-piece plastic console covers. Some of the highest priced dishwashers feature electronic touchpad controls with lighted displays for an uncluttered, high-tech look. And the highest-end European models now integrate controls on the top of the door so they can't be seen when the machine is closed.

- *Countdown timer*—This timer lets you know how much time is left in a cycle.
- *Clean light*—This light signals that the cycle is complete and the dishes are clean.
- *Soil sensors*—These take the guesswork out of cycle selection. Sensors optically analyze dirtiness of water and adjust water level and wash length accordingly.
- *Delay-start*—This timer allows starting the dishwasher automatically; it lets you take advantage of late-night, off-peak power rates or run the dishwasher after everyone has taken a shower.

Color and appearance—Does the dishwasher fit in your kitchen? Do you like its appearance?

Delay start timer—This timer allows the user to load the washer and have it start a few hours later.

Energy Efficiency

Energy Factor (EF) is the dishwasher energy performance metric. EF is expressed in cycles per kWh and is the reciprocal of the sum of the machine's electrical energy per cycle, *M*, plus the water heating energy consumption per cycle, *W*.

$$Energy\ Factor\ (EF)\ =\ \frac{1}{M + W}$$

This equation may vary based on dishwasher features such as water-heating boosters or truncated cycles. The greater the EF, the more efficient the dishwasher is. The EF is the energy performance metric of both the federal standard and the ENERGY STAR qualified dishwasher program. The federal Energy Guide label on dishwashers shows the annual energy consumption and cost. These figures use the energy factor, average cycles per year, and the average cost of energy to make the energy and cost estimates. The EF may not appear on the Energy Guide label.

Test Criteria for ENERGY STAR Qualified Dishwashers

Dishwasher manufacturers must self-test their equipment according to the new Department of Energy (DOE) test procedure defined in 10 CFR 430, Subpart B, Appendix C. This DOE test procedure was announced on August 29, 2003, and all models must be tested using the new procedure by February 25, 2004.

This test procedure establishes a separate test for soil-sensing machines. Included in the final rule was a decision to add standby energy consumption to the annual energy and cost calculation, but not to the energy factor calculation. Also, the average cycles per year have been lowered from **264 cycles** per year to **215 cycles** per year. ENERGY STAR dishwashers are at least 25 percent more energy efficient than minimum federal government standards. Table 5.3 lists the standard and the ENERGY STAR approved dishwasher energy factors.

The current ENERGY STAR criterion for dishwashers became effective January 1, 2001. This criterion of at least 25 percent above the federal standard applies only to models manufactured after January 1, 2001. The previous ENERGY STAR criterion was 13 percent above the federal standard.

Dishwashers: Your "Power" in the Environmental Protection

Buying the correct size appliance for your needs is critical to saving money, energy, and water. In dishwashers, there are compact and standard-capacity units. Compact models use less energy and water per load, so you may actually consume more energy operating them more frequently. The following tips help you to save even more:

🖙 Avoid rinsing dishes before you load them in the dishwasher or, if you must rinse, use cold water.

Table 5.3   *Energy Standards for Dishwashers*

PRODUCT TYPE	FEDERAL STANDARD ENERGY FACTOR	ENERGY STAR ENERGY FACTOR
Standard (≥ 8 place settings + six serving pieces)	≥ 0.46	≥ 0.58
Compact (< 8 place settings + six serving pieces)	≥ 0.62	NA

- Run your dishwasher with a full load. Most of the energy used by a dishwasher goes to heat water. Since you can't decrease the amount of water used per cycle, fill your dishwasher to get the most from the energy used to run it.

- Avoid using the heat-dry, rinse-hold, and pre-rinse features. Instead use your dishwasher's air-dry option. If your dishwasher does not have an air-dry option, prop the door open after the final rinse to dry the dishes.

Sources

Water Heaters and Energy Conservation—Choices, Choices, *Home Energy Magazine Online,* May/June 1996.

http://www.aceee.org/consumerguide/topwater.htm

http://www.consumerenergycenter.org/homeandwork/homes/inside/appliances/water heaters.html

http://www.eere.energy.gov/femp/pdfs/electric_waterheater.pdf

http://www.eere.energy.gov/buildings/info/components/waterheating/solarhot.html

http://www.eere.energy.gov/buildings/info/components/waterheating/heatpump.html

http://www.pueblo.gsa.gov/cic_text/housing/H2Oheater/nwtrhtr.pdf

http://www.healthgoods.com/Education/Healthy_Home_Information/Home_Appliances/ refrigerators.htm

http://www.energystar.gov/ia/partners/manuf_res/bernstudy.pdf

http://www.eere.energy.gov/consumerinfo/energy_savers/virtualhome/508/dish washer.html

questions

1. Describe three operating practices in using refrigerators that can save energy and money.

2. List the five main components of a refrigerator and explain how a refrigerator works.

3. Describe five ways in which you can reduce the energy consumption of water heaters at home with good operating practices.

4. Explain the advantages and disadvantages of Storage and Demand water heaters.

5. What are the various methods in which solar energy can be collected for water heating?

6. What is EF? And how does it describe the efficiency of a water heater?

7. How do electricity and gas compare for water heating?

8. Why are ENERGY STAR appliances better in terms of efficiency, and how can they help the environment?

9. What are Energy Guide labels? What information can be obtained from these labels, and how can this information be used to select an environmentally friendly appliance?

10. Briefly describe five ways in which we can save energy using clothes washers and dryers.

11. Compare and contrast the v-axis (top loading) and h-axis (front loading) water heaters.

12. Explain how the energy efficiency of clothes washers is evaluated.

13. What are good operating practices for clothes dryers?

14. How can you describe the energy efficiency of dishwashers?

15. Describe appliance-operating practices that can help the environment.

multiple choice questions

1. How can you reduce your water heating bill (which accounts for about 14 percent of the average utility bill)?
   a. Use less hot water
   b. Turn down the thermostat on your water heater
   c. Insulate your water heater
   d. Buy a new, more efficient water heater
   e. All of the above

2. For energy-efficient operation of a dishwasher
   a. Wash only full loads
   b. Use the no-heat air-dry feature
   c. Don't pre-rinse dishes before putting them in the dishwasher
   d. All of the above
   e. None of the above

3. The most efficient washer is
   a. Front loading
   b. Side loading
   c. Top loading
   d. Bottom loading

4. Most energy efficient appliances have an
   a. Energy Guide label
   b. ENERGY STAR label
   c. Energy Efficiency label

5. Water heaters are typically set to
   a. 90°F
   b. 120°F
   c. 150°F
   d. 180°F

6. On-demand water heaters have a storage tank.
   a. True
   b. False

7. Which of the following actions is not likely to improve the efficiency of a refrigerator?
   a. Cleaning the coils in the back of the refrigerator.
   b. Checking the seals using the "dollar-bill test."
   c. Switching the motor from a DC motor to a more efficient AC motor.
   d. Keeping the freezer as full as possible.

8. All appliances have an
   a. Energy Guide label
   b. ENERGY STAR label
   c. Energy Efficiency label

9. Natural gas water heaters are cheaper to install than electric water heaters.
   a. True
   b. False

10. Which term is used to define the energy efficiency of a dishwasher?
    a. AFUE
    b. EER
    c. SEER
    d. EF

problems for practice

1. Calculate the amount of heat energy required to heat 200 lbs of water when heated from 55°F to 130°F.

2. A family of four heats 200 gallons of water in a water heater from 60°F to 120°F every day. What is the annual energy requirement?

3. An electric water heater heats 250 gallons of water per day from 58°F to 140°F. How many kWh of energy are required? Recall that 3,412 Btus = 1 kWh.

4. If the temperature of the water heater was reduced to 120°F in the previous question, what percent of energy can be saved?

5. What is the cost of operating the water heater in Problem 3 if electricity cost is $0.08 per kWh?

6. A water heater heats 200 gal of water a day from 55°F to 130°F using natural gas. How many CCF of natural gas are required every month? (1 CCF of gas has 100,000 BTUs.)

7. What would be the monthly cost of natural gas in Problem 6? Assume a price of $10/MMBTUs for natural gas.

8. What would be the monthly cost of energy if electricity was used for heating the water in Problem 6? Assume a price of $0.08/kWh for electricity (3,412 BTUs = 1 kWh).

9. If the temperature of the water heater was reduced to 120°F in Problem 6 and electricity is used for heating, how many kWh could be saved and what would be the cost savings?

10. Estimate the percent energy savings of an electric water heater that heats 100 gallons of water per day when the temperature is set back at 110° instead of 120°F. The basement is heated and is at 65°F. The life of the water heater is expected to be about 15 years. Use an appropriate cost for electricity and compare the operating expenses with the approximate initial cost of the water heater (from the lectures).

$$\text{Heat required (BTUs)} = m \times C_p \times \text{(Temperature difference)}$$

where C_p is the heat capacity of water (1 Btu/lb/F) and m is the mass of the water.

(Assume 1 gal has 8.3 lb of water and that 3,412 BTUs = 1 kWh.)

11. An old refrigerator consumed 150 W of power. Assuming that the refrigerator operates for 20 hours in a day, what is the annual operating cost assuming the cost of electricity to be $0.06 per kWh?

12. Suppose you are comparing two ovens, both of which last for 10 years. The least efficient oven draws 1,000 W. The most efficient one uses 450 W. Assuming that the oven uses 700 h annually and that the local energy costs $0.06 per kWh, can you save any money? If so, how much money over its lifetime? If the more efficient oven costs $100 more than the least efficient one, would you buy the more efficient model?

13. Suppose you are comparing two refrigerators, both of which last for 10 years. The least efficient refrigerator draws 275 W of power. The most efficient one uses 250 W. Assuming that the refrigerator operates 4,000 hours annually and that the local energy costs $0.06 per kWh, can you save any money with the energy efficient model? If so, how much money over its lifetime? If the more efficient refrigerator costs $100 more than the least efficient one, would you buy the more efficient model?

$\mathcal{L}$ighting

goals

- To explain basic lighting principles and definitions

- To know the various types of lighting and to explain how each type works

- To explore energy efficient lighting options

- To determine life cycle costs(LCC) for different types of lighting

*L*ighting accounts for 20 to 25 percent of all electricity consumed in the United States. An average household dedicates 5 to 10 percent of its energy budget to lighting, while commercial establishments consume 20 percent to 30 percent of their total energy just for lighting.

In a typical residential or commercial lighting installation, 50 percent or more of the energy is wasted by obsolete equipment, inadequate maintenance, or inefficient use. Technologies developed during the past 15 years can help us cut lighting costs 30 percent to 60 percent while enhancing lighting quality and reducing environmental impacts.

How Is Lighting Measured?

When most people buy a light bulb they look for watts (W). Recall that watt is a unit of power, i.e., the rate at which energy is consumed from the electricity supplier. It does not say anything about the light. The most common measure of light output (or luminous flux) is the lumen. All lamps are rated in lumens as shown in Figure 6.1, and every bulb has three parameters listed on the package—the lamp lumen output, power consumption in watts and the life of the bulb in hours. For example, a 40-watt bulb produces about 505 lumens and has a life of about 1,000 hours. When this bulb is used to light a room of 10×10 feet, these 505 lumens are distributed over 100 square feet of floor area.

The distribution of light on a horizontal surface is called its **illumination** and is measured in **foot-candles.** A foot-candle (fc) of illumination is a lumen of light distributed over a 1-square-foot (0.09-square-meter) area. So, when 505 lumens of light produced by a 40-watt bulb are distributed over 100 sq. ft of area of a room, each sq. ft will receive 5.05 lumens, or 5.05 fc is the illumination.

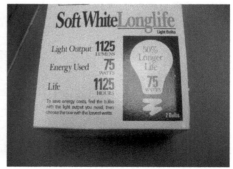

Figure 6.1   *Parameters listed on a light bulb.*

How Much Light Is Needed?

How much light is needed in a room depends on the task(s) being performed (contrast, requirements, space, size, etc.).There are three different types of tasks: ambient, task, and accent. The light requirement also depends on the ages of the occupants and the importance of speed and accuracy of the task. **Ambient lighting** is the general purpose lighting—in the hallways for safety and security. An illumination of 30 fc is generally the maximum that one needs for this purpose. **Task lighting** is lighting designed for specific tasks like cooking, sewing, or repairing a wrist watch. These tasks require more than ambient lighting and about 50–100 fc. However, the area of this level of illumination will be small. Increasing the light everywhere is not required and is a waste of energy. **Accent lighting** is the lighting that is provided to highlight certain objects or areas. For example, floodlights highlight a painting or a statue. Accent lighting also illuminates walls so they blend more closely with naturally bright areas like ceilings and windows. Accent lighting can be high intensity or subtle.

Color Rendering Index

Lamps are assigned a color temperature (according to the Kelvin temperature scale) based on their "coolness" or "warmness." The human eye perceives colors as cool if they are at the blue-green end of the color spectrum and warm if they are at the red end of the spectrum. The ability to see colors properly is another aspect of lighting quality. An object's color appears to be different under different types of light. The color rendering index (CRI) scale is used to compare the effect of a light source on the color appearance of its surroundings. A scale of 0 to 100 defines the CRI. A higher CRI means better color rendering, or less color shift.

Factors Affecting the Number of Lamps Required

- *Fixture efficiency*—Certain fixtures reflect more light than others. The fixtures that are not highly reflective may absorb some light, resulting in less light reaching the user.

- *Lamp lumen output*—The efficiency of a bulb increases with wattage. For example a 40-watt bulb produces 505 lumens whereas a 100-watt bulb (2.5 times the 40 watts) produces 1750 watts (4.32 times).

- *The reflectance of surrounding surfaces*—Bright colors or reflective surfaces painted with glossy texture finishes will appear brighter than surfaces with flat finish paint.

■ *The effects of light losses from lamp lumen depreciation and dirt accumulation—* As the lamps age or as dirt accumulates on the bulb surface, the lumen output from the light bulb decreases. Therefore, newer light bulbs produce more light than older bulbs at the same wattage.

■ Room size and shape

■ Availability of natural light (daylight)

Types of Lighting

There are four basic types of lighting:

1. Incandescent,

2. Fluorescent,

3. High-intensity discharge, and

4. Low-pressure sodium

Incandescent Bulbs

Thomas Alva Edison invented the incandescent light bulb with reasonable life. Lewis Latimer perfected the use of thecarbon filament. As shown in Figure 6.2, the incandescent bulb consists of a sealed glass bulb with a filament inside. When electricity is passed through the filament, the filament gets hot. Depending on the temperature of the filament, radiation is emitted from it. The filament's temperature is very high, generally over 2,000°C, or 3,600°F. In a "standard" 60-, 75-, or 100-watt bulb, the filament temperature is roughly

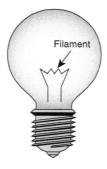

Filament

Figure 6.2   *Schematic of an incandescent bulb.*

2,550°C, or roughly 4,600°F. At high temperatures like this, the thermal radiation from the filament includes a significant amount of visible light. This principle of obtaining light from heat is called "incandescence." At this high temperature of 2,000°C, about 5 percent of the electrical energy converts into visible light and the rest of it is emitted as heat or infrared radiation.

Standard incandescents are the most common yet the most inefficient. Larger wattage bulbs have a higher efficacy than smaller wattage bulbs. Figure 6.3 shows the increase in efficiency with higher wattage bulbs.

Tungsten halogen is an incandescent lamp with gases from the halogen family sealed inside the bulb and an inner coating that reflects heat back to the filament (Figure 6.4). It has similar light output to a regular incandescent bulb but less power. Halogens in the gas filling reduce the material losses of the filament caused by evaporation and increase the performance of the lamp.

Tubular tungsten-halogen bulbs are commonly used in "torchiere" floor lamps (Figure 6.5), which reflect light off of the ceiling, providing more diffused and suitable general lighting. Although these lamps provide better energy efficiency than the standard A-type bulb, they consume significant amounts of energy (typically drawing 300 to 600 W) and become very hot (a 300-watt tubular tungsten-halogen bulb reaches a temperature of about 260°C compared to about 60°C for a compact fluorescent bulb). However, tungsten-halogen lamps operate at very high temperatures and should not be used in fixtures that have paper- or cellulose-lined sockets. A halogen bulb is often 10 to 20 percent more efficient than an ordinary incandescent bulb of similar voltage, wattage, and

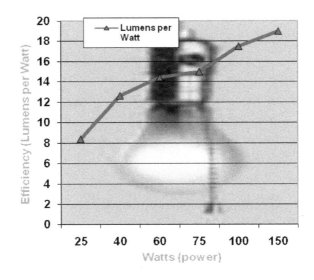

Figure 6.3   *Efficiency of an incandescent light bulb.*

Figure 6.4

A tungsten-halogen lamp.

Figure 6.5

A "torchiere" floor lamp.

life expectancy. Halogen bulbs may also have two to three times as long a lifetime as ordinary bulbs. How much the lifetime and efficiency are improved depends largely on whether a premium fill gas (usually krypton, sometimes xenon) or argon is used. Figure 6.6 shows a picture taken with an infrared camera comparing heat produced by a halogen and that by a compact fluorescent light bulb.

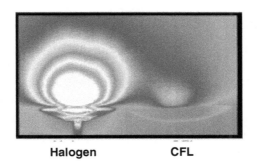

Halogen   **CFL**

Reflector Lamps

Light waves from a bulb spread in all directions. The light that goes toward the back is not useful when the light is most needed in the front. Reflector lamps (Type R) are designed to spread light over specific areas. They have silver coating on the sides like any mirror, and all the light waves passing through the sides or the back are reflected to the front. Therefore, they are called reflector lamps and are also called floodlighting, spotlighting, and down lighting bulbs. Parabolic aluminized reflector (PAR) lamps are also available with halogen technology to operate at 120 volts as shown in Figure 6.7. A standard 150-W incandescent spotlight can be replaced with a lower wattage halogen lamp, reducing electricity consumption by up to 40 percent.

Figure 6.7

A reflector lamp (Type R) light bulb.

Fluorescent Bulbs

The fluorescent bulb is a major advancement and a commercial success in small-scale lighting since the original tungsten incandescent bulb. These bulbs are highly efficient compared to the incandescent bulbs. Fluorescence is the phenomenon in which absorption of light of a given wavelength by a fluorescent molecule is followed by the emission of light at longer wavelengths.

Fluorescent lamps provide light by the following process as shown in Figure 6.8.

1. An electric discharge (current) is maintained between the electrodes through the mercury vapor and inert gas.
2. This current excites the mercury atoms, causing them to emit non-visible ultraviolet (UV) radiation.
3. This UV radiation is converted into visible light (fluorescence) by the phosphors lining the tube.

Fluorescent lights need ballasts (i.e., devices that control the electricity used by the unit) for starting and circuit protection. Ballasts consume energy. Figure 6.9 shows a picture of a full-size fluorescent lamp fixture. The energy savings for existing fluorescent lighting can be increased by re-lamping (e.g., replacing an existing lamp with one of a lower wattage), replacing ballasts, and replacing fixtures with more efficient models.

Fluorescent lamps are about 2 to 4 times more efficient than incandescent lamps at producing light at the wavelengths that are useful to humans. Thus, they run cooler for the same effective light output. The bulbs themselves also last a lot longer—10,000 to 20,000 hours versus 1,000 hours for a typical incandescent. However, for certain types of ballasts, this is only achieved if the fluorescent lamp is left on for long periods of time without frequent on-off cycles. Full-size fluorescent lamps are available in several shapes, including straight, U-shaped, and circular configurations. Lamp diameters range from 1″ to 2.5″. The most common lamp type is the four-foot (F40), 1.5″ diameter (also called T12)

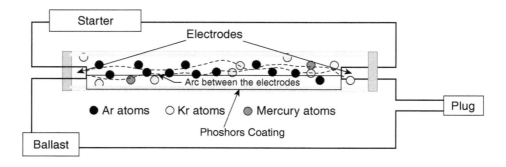

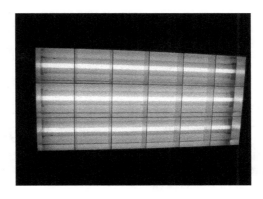

Figure 6.9

Full-size fluorescent lamp fixture.

straight fluorescent lamp. More efficient fluorescent lamps are now available in smaller diameters, including the 1.25″ (also called T10) and 1″ (also called T8).

Fluorescent lamps are available in color temperatures ranging from warm (2,700 K) "incandescent-like" colors to very cool (6,500 K) "daylight" colors. "Cool white" (4,100 K) is the most common fluorescent lamp color. Neutral white (3,500 K) is becoming popular for office and retail use.

Compact Fluorescent Lamps

These are miniaturized fluorescent lamps that usually have premium phosphors that often come packaged with integral or modular ballast, as shown in Figure 6.10. They typically have a standard screw base that can be installed into nearly any table lamp or lighting fixture that accepts an incandescent lamp, and these operate at standard voltage as incandescent bulbs.

Compact fluorescent lights (CFLs) come in a variety of sizes and shapes and are being used as energy saving alternatives to incandescent lamps. They also have a much longer life—6,000 to 20,000 hours (10 to 20 times longer) compared to 750 to 1,000 hours for a standard incandescent.

CFLs can replace incandescent bulbs that are roughly 3 to 4 times their wattage but can cost up to 10 times more than comparable incandescent bulbs

They are one of the best energy-efficiency investments available. Although they cost more they are very economical in a long run.

High-Intensity Discharge Lamps

High-intensity discharge (HID) lamps are similar to fluorescents in that an arc is generated between two electrodes. The arc in a HID source is shorter, yet it generates much more light, heat, and pressure within the arc tube.

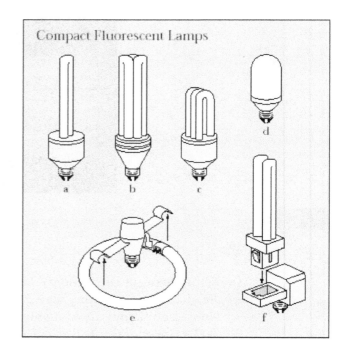

Compact fluorescent lamps (CFLs) come in a variety of sizes and shapes including (a) twin-tube integral, (b and c) triple-tube integral, (d) integral model with casing that reduces glare, (e) modular circline and ballast, and (f) modular quad-tube and ballast. CFLs can be installed in regular incandescent fixtures, and they consume less than one-third as much electricity as incandescent lamps do.

Figure 6.10

Types of compact fluorescent bulbs available in the market.
Source: *http://www.eere.energy.gov/consumerinfo/pdfs/eelight.pdf*

Originally developed for outdoor and industrial applications, HID lamps are also used in office, retail, and other indoor applications. Their color rendering characteristics have been improved, and lower wattages have recently become available (as low as 18 watts).

There are several advantages to HID sources:

- relatively long life (5,000 to 24,000+ hrs)

- relatively high lumen output per watt

- relatively small in physical size

However, the following operating limitations must also be considered. First, HID lamps require time to warm up. It varies from lamp to lamp, but the average warm-up time is 2 to 6 minutes. Second, HID lamps have a "restrike" time, meaning a momentary interruption of current or a voltage drop too low to maintain the arc will extinguish the lamp. At that point, the gases inside the lamp are too hot to ionize, and time is needed for the gases to cool and pressure to drop before the arc will restrike. This process of restriking takes between 5 and 15 minutes, depending on which HID source is being used. Therefore, good applications of HID lamps are areas where lamps are not switched on and off intermittently. They also require ballasts, and they take a few seconds to produce light when first turned on because the ballast needs time to establish the electric arc.

The following HID sources are listed in increasing order of efficacy (lumens per watt):

- mercury vapor
- metal halide
- high-pressure sodium
- low-pressure sodium

Mercury vapor lamps are widely used to light both indoor and outdoor areas such as gymnasiums, factories, department stores, banks, highways, parks, and sports fields. Mercury vapor lamps consist of an inner arc discharge tube constructed of quartz surrounded by an outer hard borasilicate glass envelope. Shortwave UV, a result of the decay of mercury atom electrons from an excited to a stable state, is readily transmitted through the inner quartz tube but is virtually blocked by the outer glass envelope during normal operation.

Metal halide lamps are similar to mercury vapor lamps but use metal halide additives inside the arc tube along with the mercury and argon. These additives enable the lamp to produce more visible light per watt with improved color rendition. Wattages range from 32 to 2,000, offering a wide range of indoor and outdoor applications. The efficacy of metal halide lamps ranges from 50 to 115 lumens per watt typically about double that of mercury vapor. In short, metal halide lamps have several advantages:

- high efficacy
- good color rendering
- wide range of wattages

However, they also have some operating limitations:

- The rated life of metal halide lamps is shorter than other HID sources; lower-wattage lamps last less than 7,500 hours while high-wattage lamps last an average of 15,000 to 20,000 hours.
- The color may vary from lamp to lamp and may shift over the life of the lamp and during dimming.

Because of the good color rendition and high lumen output, these lamps are good for sports arenas and stadiums. Indoor uses include large auditoriums and convention halls. These lamps are sometimes used for general outdoor lighting, such as parking facilities, but a high-pressure sodium system is typically a better choice.

The high-pressure sodium (HPS) lamp is widely used for outdoor and industrial applications. Its higher efficacy makes it a better choice than metal halide for these applications, especially when good color rendering is not a priority.

HPS lamps differ from mercury and metal-halide lamps in that they do not contain starting electrodes; the ballast circuit includes a high-voltage electronic starter. The arc tube is made of a ceramic material that can withstand temperatures up to 2,372°F. It is filled with xenon to help start the arc, as well as a sodium-mercury gas mixture.

The efficacy of the lamp is very high (as much as 140 lumens per watt). For example, a 400-watt high pressure sodium lamp produces 50,000 initial lumens. The same wattage metal halide lamp produces 40,000 initial lumens, and the 400-watt mercury vapor lamp produces only 21,000 initially.

Sodium, the major element used, produces the "golden" color that is characteristic of HPS lamps. Although HPS lamps are not generally recommended for applications where color rendering is critical, HPS color rendering properties are being improved. Some HPS lamps are now available in "deluxe" and "white" colors that provide higher color temperature and improved color rendition. The efficacy of low-wattage "white" HPS lamps is lower than that of metal halide lamps (lumens per watt of low-wattage metal halide is 75–85, while white HPS is 50-60 LPW).

Low-pressure sodium lamps—producing up to 180 lumens per watt—have the highest efficacy of all commercially available lighting sources. A low-pressure sodium lamp is shown in Figure 6.11.

Even though they emit a yellow light, a low-pressure sodium lamp shouldn't be confused with a standard high-pressure sodium lamp—a high-intensity discharge lamp (Figure 6.12). Low-pressure sodium lamps operate much like a fluorescent lamp and require ballast. The lamps are also physically large—about 4-feet long for the 180-watt size—so light distribution from fixtures is less controllable. There is a brief warm-up period for the lamp to reach full brightness.

With a CRI of 0, low-pressure sodium lamps are used where color rendition is not important but energy efficiency is. They're commonly used for outdoor, roadway, parking lot, and pathway lighting. Low-pressure sodium lamps are

Figure 6.11

A low-pressure sodium lamp.

Figure 6.12

A high-pressure (mercury) vapor lamp.

preferred around astronomical observatories because the yellow light can be fil-tered out of the random light surrounding the telescope.

Life Cycle Cost Analysis

Performing a life-cycle cost analysis (LCC) gives the total cost of a lighting sys-tem—including all expenses incurred over the life of the system. This analysis can be applied not only to lighting but for most of the appliances, automobiles, heating systems, and so on when two systems are compared to determine cost effective option. There are two reasons to do an LCC analysis: (1) to compare different systems or bulbs in this case, and (2) to determine the most cost-effec-tive system or bulb. For some lighting systems the initial cost may be high but the energy costs will be low over their lifetime. In other cases, the initial cost to buy a bulb or a system may be low, along with the energy or maintenance costs, but the useful life of such a bulb or system may be short. In that case we may have to replace the appliance several times to get the same useful life as the other option. Therefore, a life-cycle cost (LCC) analysis can be helpful for comparing the total costs incurred over its lifetime. It is in essence calculating all the costs incurred to buy, maintain, and run over its lifetime.

Life Cycle Costs = Cost to buy + cost to maintain it (if any maintenance is required) + cost of energy to run it for its life + replacement costs – any salvage value

Where:

Cost to buy = the purchase price of the bulb or the system.

Cost to maintain is the cost incurred to maintain it in good operating condition. For example, in the case of a car, engine oil change every 3,000 miles is part of maintenance costs.

Cost of energy is the energy or the fuel it takes to run the appliance or bulb for its lifetime.

Replacement Cost

If bulb A has a life of 1,000 hours and bulb B has a life of 10,000 hours, then bulb A needs to be replaced 10 times to get the same useful life period as that of bulb B. Therefore, 10 A bulbs need to be purchased for each B bulb.

Table 6.1 shows a life cycle cost analysis in comparing an incandescent bulb and a CFL.

This analysis shows clearly that each incandescent bulb replaced with a CFL that would fit into a regular incandescent bulb fixture would save about $39.25 over 10,000 hours of operation. Imagine the number of bulbs that you have at home: living room, kitchen, bathrooms, bedrooms, table lights, floor lamps, ceiling lights, closets, garage, basement, and so on. There are energy saving options all over the home.

Table 6.1   *Life Cycle Cost of a Light Bulb*

	INCANDESCENT	COMPACT FLUORESCENT LIGHT (CFL)
Rating	60 Watts	15 Watts
Lumen output	865 lumens	900 lumens
Cost to buy the bulb ($)	$0.60	$5.00
Life of each bulb	1,000 h	10,000 h
Bulbs needed for same life	10 bulbs—$6.00	1 bulb—$5.00
Energy consumption	60 Watts × 10,000 h 600,000 Wh = 600 kWh	15 Watts × 10,000 h 150,000 Wh = 150 kWh
Price of electricity/kWh	$0.085	$0.085
Cost of electricity needed for 10,000 h	600 kWh × 0.085/kWh $51.00	150 kWh × 0.085/kWh $12.75
Total cost (life cycle costs) to own and operate the bulbs for 10,000 h	$51.00 + $6.00 **$57.00**	$12.75 + $5.00 **$17.75**

However, there are some disadvantages with CFLs.

- They are often physically larger than the incandescent bulbs they replace and simply may not fit the lamp.
- The light is generally cooler—less yellow—than incandescent light bulbs. This may result in less than pleasing contrast with ordinary lamps and ceiling fixtures. Newer models have addressed this issue.
- Some types (usually iron ballasts) may produce an annoying flicker.
- Ordinary dimmers cannot be used with normal compact fluorescent bulbs unless they have dimmable ballast.
- Like other fluorescent bulbs, operation at cold temperatures (under around 50–60°F) may result in reduced light output.
- There may be an audible buzz from the ballast.

Efficacy of Light Bulbs

This is the ratio of light output from a lamp to the electric power it consumes and is measured in lumens per watt (LPW). Figure 6.13 shows the efficacy of various types of lighting. It can be seen from the figure that there are very efficient lighting options available for most residential, commercial, and industrial applications.

LED Lighting

A typical signalized four-way intersection may have three lanes per approach (two through lanes and a left turn lane), plus pedestrian crossings (Figure 6.14). Conventional incandescent lamps in a single four-way traffic light consume roughly 85 kWh of electricity per day and cost about $1,600 per year to operate. LED lights use just 10 percent of the electricity that incandescent lamps use, so the opportunity for savings is enormous in this country.

> ### Did you know?
> *Four million traffic lights in the U.S. consume about 3 billion kWh of electricity.*
> http://www.solaraccess.com/news/story?storyid=880 From the 9/10/2001 Edition

Improved Lighting Controls for Energy Efficiency

Lighting controls give you the flexibility to design a space for multiple use and easy access. They should be a part of the lighting plan for every room. Both manual and automatic controls can cut energy costs by making it easier to use lights

Light Output (lumens per watt)

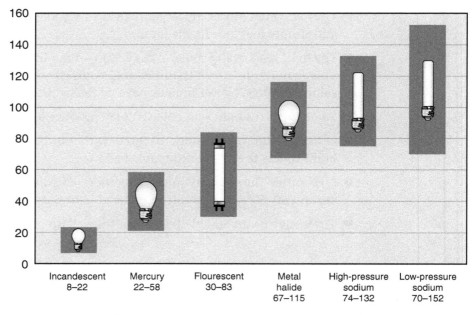

Each type of lamp differs in the amount of light it delivers per watt.

Efficacy of various types of lighting.

LED traffic lights.
Source: *http://www.scottsdaleaz.gov/Traffic/SignsSignals/led.asp*

only when and where they are needed. Controls used with high-wattage incandescent bulbs are especially effective for saving energy, but they should be considered for use with any lights that might be left on when no one is using them. Always choose controls that are compatible with the bulb and ballast. Try to obtain the best quality so the controls will perform well over time.

Switches

The simple on-off switch, whether mounted on a wall or on the light fixture, should always be obvious and convenient. On fixtures with pull-cord switches, attach an object at the end that is easy to see and grasp. It's always a good idea to install multiple wall switches in areas that have more than one entrance, such as hallways, staircases, and large rooms. These switches should be easy to find as well. It's a simple matter to add devices such as oversize toggles or switch plates that glow in the dark. A small indicator light near a switch can signal when lights that are out of sight—in the basement or outdoors, for example—have been left on.

If a main switch in a room controls several lights, each fixture should have its own switch so individual lights can be turned off if they are not needed. In rooms such as kitchens, where lights are used for different purposes, overhead ambient lights, counter, or island lights should be on separate switches. A three-level switch in lamps is a simple way to use one fixture for several lighting needs. When the higher levels are not necessary, switch to the lowest level to save energy.

Photocells

A photosensor measures the light level in an area and turns on an electric light when that level drops below a set minimum. They are most effective with lights that stay on all night long, such as some outdoor fixtures or night lights. If a light does not need to remain on throughout the night, use a timer or motion detector.

Timers

Timers are an inexpensive way to control the amount of time a light stays on inside the home or outdoors. They can be located at a light switch, at a plug, or in a socket. Some models are turned on manually and set to turn off after a designated number of minutes or hours. Others can be programmed to turn on and off at specified times. Both mechanical and solid-state timers are available, and some offer the option of a manual override. Some screw-base compact fluorescent bulbs cannot be used with timers, so check the manufacturer's recommendations.

Be careful not to set timers so a light might turn off in an area when someone could be left in the dark. Or install a glow-in-the-dark switch plate or a very low-wattage nightlight with a photosensor near the switch so it is easy to find.

Motion or Occupancy Sensors

Motion detectors, or occupancy sensors, have proven to be an excellent way to save energy, especially in bathrooms and bedrooms where lights are frequently left on. They are also popular outdoors for walkways, driveways, and as security lights.

Sensors can operate automatically to turn lights on when movement is detected, then off after a specified period of no motion, or they can have manual on or off switches. Some models feature dimmers that reduce light to a preset level rather than turn completely off when there is no movement; others come with photo sensors that turn lights on only when the light level is below a preset point and motion is detected. Follow the manufacturer's instructions for installing sensors to ensure the proper coverage area. Also be sure the lights are compatible with the sensors. Some compact fluorescents should not be used with motion detectors, nor should high intensity discharge lights because of their inability to relight quickly.

Dimmers

Dimming fluorescent lamps is not all that easy to do. If you reduce power to the lamp, the filaments will not be as hot and will not be able to thermionically emit electrons as easily. If the filaments get too cool by dimming the lamp greatly, usually the lamp will just go out. If you force current to continue flowing while the electrodes are at an improper temperature, then severe rapid degradation of the thermionic material on the filaments is likely. To effectively, reliably, and safely dim fluorescent lamps below around half brightness or so, you need special equipment that may only work properly with a specific lamp. Such equipment typically gives some power to the filaments to keep them at a workable temperature while the current flowing through the bulb is greatly reduced.

Manual dimming controls allow occupants of a space to adjust the light output or illumination. This can result in energy savings through reductions in input power, as well as reductions in peak power demand, and enhanced lighting flexibility.

Fluorescent lighting fixtures require special dimming ballasts and compatible control devices. Some dimming systems for high-intensity discharge lamps also require special dimming ballasts. Table 6.2 provides a comparison of characteristics of various types of bulbs.

Sources

http://www.eere.energy.gov/EE/buildings_lighting.html
http://www.ge.com/en/product/home/lighting.htm
http://members.misty.com/don/light.html
http://hes.lbl.gov/hes/makingithappen/no_regrets/lightingcontrols.html
http://www.eere.energy.gov/buildings/components/lighting/lamps/lowpressure.cfm

Table 6.2   *Characteristics of Various Types of Light Bulbs*

INCANDESCENT	FLUORESCENT	HIGH INTENSITY DISCHARGE (HID)
Does not require a ballast	Requires a ballast	Requires a ballast
Warm color appearance with a low color temperature and excellent color rendering (CRI 100)	Range of color temperatures and color rendering capabilities	Ambient temperature does not affect light output, although low ambient temperatures can affect starting, requiring a special ballast
Compact light source	Low surface brightness compared to point sources	Compact light source
Simple maintenance due to screw-in Edison base	Cooler operation	High lumen packages
Less efficacy light source	More efficacious compared to incandescent	Point light source
Shorter service life than other light sources in most cases	Ambient temperatures and convection currents can affect light output and life	Range of color temperatures and color rendering abilities depending on the lamp type
Filament is sensitive to vibrations and jarring	All fixtures installed indoors must use a Class P ballast that disconnects the ballast in the event it begins to overheat; high ballast operating temperatures can shorten ballast life	Long service life
Bulb can get very hot during operation		Highly efficacious in many cases
Must be properly shielded because incandescent lamps can produce direct glare as a point source	Options for starting methods and lamp current loadings	Line voltage variations, possible line voltage drops, and circuits sized for high starting current requirements must be considered
Requires proper line voltage as line voltage variations can severely affect light output and service life	Requires compatibility with ballast	
	Low temperatures can affect starting unless a "cold weather" ballast is specified	

questions

1. How is light measured?

2. What factors determine the amount of light that is needed in a room?

3. What are the three main methods of producing light?

4. Explain the difference between incandescence, fluorescence, and high-intensity discharge.

5. What are common ways in which we can improve the energy efficiency?

multiple choice questions

1. Approximately what percentage of electricity does an incandescent light bulb convert into visible light?
   a. 5
   c. 40
   b. 20
   d. 90

2. Light is measured in these units.
   a. Amperes
   c. Watts
   b. Henrys
   d. Lumens

3. A 100-watt incandescent light bulb is operated for 12 hours, and a 15-watt fluorescent light bulb is operated for the same period of time. At 10 cents per kWh, what is the cost savings of the fluorescent bulb?
   a. $1.02
   b. $0.10
   c. $0.01

4. This amount of lighting is required for reading.
   a. 20 fc
   c. 90 fc
   b. 150 fc
   d. 50 fc

5. Compact fluorescent lamps are how many times more efficient than incandescent lamps producing an equivalent light output?
   a. 50 percent more efficient
   b. Twice as efficient
   c. 3–4 times more efficient
   d. 10 times more efficient

6. The luminous efficacy of a tungsten lamp is about 12 lumens/watt. What is the efficacy of a fluorescent tube (in lumens/watt)?
   a. 20
   c. 80
   b. 50
   d. 100

7. A higher CRI means
   a. Better color rendering
   b. Worse color rendering
   c. CRI has nothing to do with color rendering
   d. Better lighting efficiency

8. Fluorescent bulbs are filled with
   a. Air
   b. Argon
   c. A mixture of helium and mercury
   d. A mixture of argon-krypton gas and a small amount of mercury

9. This method of lighting is the least efficient.
   a. Incandescent
   b. Fluorescent
   c. High-intensity discharge
   d. Low-pressure sodium

10. Higher-wattage bulbs are less efficient than lower-wattage bulbs.
    a. True
    b. False

11. Fluorescent light is produced by
    a. Heating a tungsten wire
    b. Converting UV to visible light
    c. Passing current through mercury vapor
    d. Passing a current through sodium vapor

12. This lighting method is the most efficient.
    a. Incandescent
    b. Fluorescent
    c. High-intensity discharge

13. A ballast is required for the following bulbs
    a. Incandescent   c. Reflector lamp
    b. Fluorescent

14. An example of high-intensity discharge lamps is
    a. Mercury vapor lamp
    b. High-pressure sodium
    c. Incandescent

15. Filaments in fluorescent bulbs are made of tungsten.
    a. True   b. False

16. Efficacy of lighting is measured by
    a. Watts per sq. ft
    b. Lumens per sq. ft
    c. Lumens per watt
    d. Lumens per dollar

17. Use of these devices for lighting reduces electricity consumption.
    a. Dimmers   c. Starters
    b. Ballasts   d. Locks

18. This bulb is usually filled with a mixture of inert gases and mercury.
    a. Incandescent   c. Halogen
    b. Fluorescent

19. General purpose lighting in hallways is also called
    a. Ambient   c. Accent
    b. Task

20. Vapor of this element fills energy-efficient bulbs.
    a. Potassium   c. Hydrogen
    b. Mercury   d. Halogen

21. CFL lights operate at the same voltage as standard incandescent lights.
    a. True
    b. False

22. Incandescent lights operate at the same wattage as fluorescent lights to give out the same amount of light.
    a. True   b. False

23. You may reduce your lighting energy costs by this much by utilizing modern technology.
    a. 90 percent   c. 75 percent
    b. 50 percent   d. 80 percent

24. A spotlight is likely to be of this type.
    a. Standard incandescent
    b. Tungsten halogen
    c. Type-R

25. Experts divide lighting uses into three categories.
    a. True   b. False

26. A foot-candle (fc) is defined as
    a. a lumen of light distributed over 1 square meter area
    b. a lumen of light distributed over 1 square foot area
    c. a joule of energy distributed over 1 square foot area
    d. a watt of power over 100 square feet of area

27. Fluorescent bulbs give out UV light.
    a. True   b. False

28. Daylighting typically has a CRI value of 100.
    a. True   b. False

29. Lighting labels must contain the following information
    a. Light output
    b. Energy used
    c. Life of the bulb
    d. All the above
    e. None of the above

problems for practice

1. A 60-watt light bulb produces 3 watts of radiant energy and 57 watts of heat energy. What is its efficiency?

2. A 100-watt light bulb is left on all day (24 hours). How much did it cost to operate the light bulb if electricity costs 5 cents per kWh?

3. A 100-watt incandescent light bulb is operated for 12 hours, and a 15-watt fluorescent light bulb is operated for the same period of time. At 10 cents per kWh, what is the cost savings of the fluorescent bulb?

4. Jackie Smith, who is very conscientious about the environment, would like to know how much energy she can save by switching to a fluorescent bulb. Estimate the total energy savings per light bulb fixture for Jackie by comparing the total costs to own and operate a 23-watt CFL bulb instead of the 100-watt incandescent bulb that she has been using. The expected life of incandescent and CFL bulbs is 1,000 h and 8,000 hours, respectively. The purchase price of an incandescent bulb is $0.50 and the CFL is $7.50. Assume that the electricity costs $0.085/kWh.

5. If Jackie Smith in problem 4 replaces 24 bulbs at home with CFLs, what would be her savings if the electricity cost is $0.085 per kWh?

Lighting Puzzle

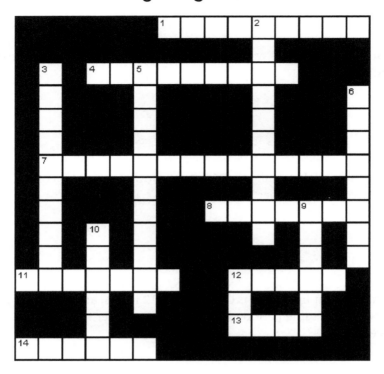

Across

1. This coating converts UV light into visible light
4. Designed to spread light over specific areas
7. This index is used to compare the effect of a light source on the color appearance of its surroundings
8. Devices that control the electricity used by a fluorescent bulb
11. This task requires the maximum amount of light
12. Measure of light output from a bulb
13. Unit of power in SI units
14. Use of this device reduces energy consumption in light bulbs

Down

2. These sense the daylight
3. A lumen of light over one square foot of area
5. This bulb is usually filled with a mixture of inert gas and mercury
6. Filaments in incandescent bulbs are made of this element
9. This type of lighting is used to highlight an object
10. Vapor of this element is filled at high pressures in efficient bulbs
12. Efficacy of a bulb is also expressed like this

*H*ome Heating Basics

goals

- *To understand the mechanisms of heat transfer*

- *To be able to calculate heating degree days for a heating season and articulate the significance of Heating Degree Day (HDD)*

- *To be able to calculate heat loss from a home using a conduction equation*

- *To understand the concept of R-value and its importance in home heating*

- *To be able to compare the cost of various fuels for a given heat loss*

- *To understand the significance and be able to calculate the payback period*

*H*ome heating is the single highest energy expense for a household. Energy spent for residential space heating accounts for about 10 percent of the total energy consumed in the United States. The average household consumes 92 million BTUs, and 46 percent of that is used for home heating (Figure 7.1). In 2001, average annual home heating expenses were $1,488. Home energy use also accounted for 20 percent of the greenhouse gas emissions. Therefore, reduction of energy consumption in home heating results in substantial monetary savings and reduction in air pollution. With appropriate improvements, average home heating costs can be reduced by 30 percent (i.e., about $500 a year) for the rest of the home's life.

Houses are heated to keep the temperature inside at about 65°F when the outside temperature is lower. A house requires heat continuously because of the heat loss. Heat can escape from a house through various places; some are well known, and some are not noticeable. Heat can escape from the roof, walls, doors, windows, basement walls, chimney, vents, and even the floor (Figure 7.2). The more heat the house leaks, the more the furnace has to put out to make up for the loss. For the furnace to generate more heat to compensate for the heat loss, more fuel needs to be put into the furnace, hence higher fuel or heating costs. It is also important to note that furnaces are not 100 percent efficient. When the furnace efficiency is lower, the fuel consumption for the same amount of heat output will be even higher.

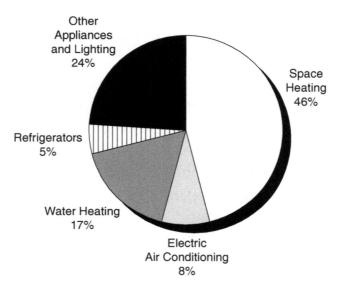

Figure 7.1

Distribution of residential energy consumption.

Figure 7.2

Air leaks in a typical house.
SOURCE: *http://www.energystar.gov/ia/home_improvement/images/houseleaks-without-textv2.gif*

Mechanisms of Heat Loss or Transfer

Heat escapes (or transfers) from inside to outside (high temperature to low temperature) by three mechanisms (either individually or in combination) from a home: (1) conduction, (2) convection, and (3) radiation (Figure 7.3).

Conduction is a process by which energy is conducted down the rod as the vibrations of one molecule are passed to the next, but there is no movement of energetic atoms or molecules (Figure 7.4). In solids the atoms or molecules do not have the freedom to move, as in liquids or gases, so the energy is stored in the vibration of atoms. An atom or molecule with more energy transfers energy to an adjacent atom or molecule by physical contact or collision. Heat is transferred by conduction through solids like walls, floors, and the roof.

Convection is a process by which energy is transferred by the bulk motion of the fluids (gases or liquids). This is the mechanism by which heat is lost by warm air leaking to the outside when the doors are opened or cold air leaking into the house through the cracks or openings in walls, windows, or doors (Figure 7.5). When cold air comes in contact with the heater in a room, it absorbs the heat and rises. Cold air, being heavy, sinks to the floor and gets heated, and thus it slowly heats the whole room.

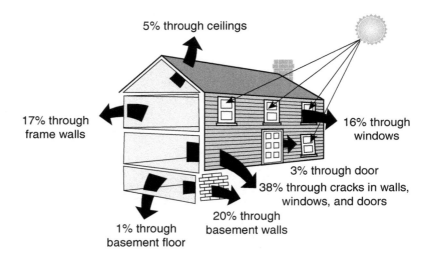

5% through ceilings

17% through
frame walls

16% through
windows

3% through door

38% through cracks in walls,
windows, and doors

20% through
basement walls

1% through
basement floor

Figure 7.3   *Typical heat loss through a house.*

heat

Figure 7.4   *An example of heat transfer by conduction.*

heat

Figure 7.5   *An example of heat transfer by convection.*

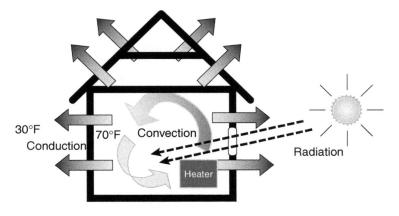

Figure 7.6

Examples of heat transfer by conduction, convection, and radiation.

Radiation is the transfer of heat through electromagnetic radiation. In gases, liquids, and solids, energy is transferred by the molecules with or without their physical movement (convection or conduction). Radiation does not need any medium (molecules or atoms). Energy can be transferred by radiation even in a vacuum (Figure 7.6).

There are two ways in which we can reduce energy consumption. The first and the most cost-effective way is to improve the home's "envelope"—the walls, windows, doors, roof, and floors that enclose the home—by improving the insulation (conduction losses) and sealing the air leaks (convection losses). The second way to reduce the energy consumption is by improving the efficiency of the furnace that provides the heat.

Conduction Heat Losses

Most of the heat lost through the walls is by conduction (Figure 7.7). The amount of loss varies with the size of the house (the area through which the heat can escape), the local weather or climatic conditions, and the wall's capacity to resist heat loss (R-value). The inside temperature is often constant at a comfortable temperature of 65°F. As the outside temperature falls lower than 65°F the heat is lost to the outside; the higher the temperature difference, the higher the heat loss to the outside. Insulation is rated in terms of thermal resistance, called the R-value, which indicates the resistance to heat flow. The higher the R-value, the greater is the insulating effectiveness. The R-value of thermal insulation depends on the type of material, its thickness, and its density.

$$Heat\ Loss\left(\frac{BTUs}{h}\right) = \frac{Area\ (ft^{2)}) \times Temperature\ Difference\ (°F)}{R\text{-}Value\left(\frac{ft^{2}\ °F\ h}{BTU}\right)} \qquad (7.1)$$

chapter 7   *Home Heating Basics*   **161**

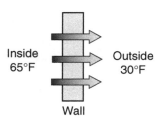

Figure 7.7

Heat loss across a solid wall by conduction.

From Equation 7.1 it can be seen that once the house is built the area of the walls will not change, and the R-value of the walls is also constant. The only variable is the temperature difference between inside and outside.

Heating Degree Days

The inside temperature is usually taken as a standard comfort temperature of 65°F. The outside temperature varies by the hour. Therefore, an average of the maximum and minimum during a day is taken as the average outside tempera-ture for an entire day. The temperature difference through which the air has to be heated is given by the Heating Degree Day (HDD). The HDD indicates how many degrees the mean temperature fell below 65°F for the day. It is also an index of fuel consumption. To calculate the heating degree days for a particular day, find the day's average outside temperature (by adding the high and low temperatures), then divide by two:

$$HDD = T_{base} - T_a \qquad (7.2)$$

where T_{base} = Base temperature or inside temperature, usually 65°F
T_a = Average outside temperature = $T_{max} + T_{min}/2$

For example, if the day's high temperature is 60°F and the low is 40°F, the aver-age temperature is 50 degrees. Therefore, HDD = 65°F – 50°F = 15°F.

If the T_a is equal to or above 65°F, there are no heating degree days for that 24-hour period, or HDD = 0.

The temperature outside may not remain the same the next day and the day after; therefore, the heat loss varies by the day. To obtain the heat loss for a whole year, we can calculate heat loss for each day and add the heat loss for all days in a year that needed heating. The R-value of the wall or the area of the wall will not change, but only the temperature outside, and therefore the difference between inside and outside will change for each day. If the average outside temperature

Illustration 7.1

For a 15-ft by 15-ft room with an 8-ft ceiling, with all surfaces insulated to R13, with inside temperature 65°F and outside temperature 25°F:

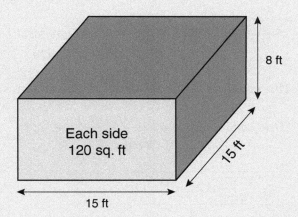

Each side
120 sq. ft

8 ft

15 ft

15 ft

$$\text{Heat loss rate} = \frac{Q(BTUs)}{t(hour)} = \frac{(480\ ft^2) \times (65°F - 25°F)}{13 \dfrac{ft^2\ °F\ h}{BTU}} = 1{,}477 \frac{BTUs}{h}$$

In a 24-hour period or one day, the heat loss would be:

$$\text{Heat loss per day} = \frac{1{,}477\ BTUs}{\cancel{h}} \times \frac{24\ \cancel{h}}{day} = 35{,}466 \frac{BTUs}{day}$$

were 35°F, 32°F, 28°F, and so on for each day, the heat loss for the whole year or the season could be calculated as follows:

$$Heat\ Loss = \underbrace{\frac{480\ ft^2 \times (65 - \overbrace{35}^{outside\ temp\ for\ day\ 1})°F}{13 \dfrac{ft^2\ °F\ BTU}{h}} \times 24 \frac{h}{day}}_{Day\ 1\ Heat\ Loss} + \underbrace{\frac{480\ ft^2 \times (65 - \overbrace{32}^{outside\ temp\ for\ day\ 2})°F}{13 \dfrac{ft^2\ °F\ BTU}{h}} \times 24 \frac{h}{day}}_{Day\ 2\ Heat\ Loss}$$

$$+ \underbrace{\frac{480\ ft^2 \times (65 - \overbrace{28}^{outside\ temp\ for\ day\ 3})°F}{13 \dfrac{ft^2\ °F\ BTU}{h}} \times 24 \frac{h}{day}}_{Day\ 3\ Heat\ Loss} \dots and\ so\ on\ for\ all\ heating\ days$$

Since the area (480 ft²), R-value (13), and 24 h in a day are common for all heating days, we can bring those out and rewrite the equation as:

$$= \frac{480\ ft^2 \times}{13\frac{ft^2 \circ F h}{BTU}} \times 24\frac{h}{day}\left\{(65-35)^\circ F + (65-32)^\circ F + (65-28)^\circ F + \ldots \text{ and so on for all heating days}\right.$$

$$= \frac{480\ ft^2 \times}{13\frac{ft^2 \circ F h}{BTU}} \times 24\frac{h}{day}\underbrace{\left\{30^\circ F + 33^\circ F + 37^\circ F + \ldots \text{ and so on for all heating days}\right\}}_{\text{This is sum of Heating Degree Days}}$$

This equation can be rewritten in general terms as:

$$Heat\ Loss\ in\ a\ Season = \frac{Area\ (ft^2)}{R\text{-}value\left(\frac{ft^2 \circ F\ h}{BTU}\right)} \times 24\frac{h}{day} \times (HDD\ for\ the\ Season) \quad (7.3)$$

$$= \boxed{\frac{Area}{R\text{-}value} \times 24 \times HDD}$$

It is important to remember that the Equation 7.3 gives the annual or seasonal heat loss, whereas Equation 7.1 gives the heat loss per hour. Seasonal Heating Degree Days (HDD) is the sum of temperature differences of all days during which heating is required.

Illustration 7.2

Calculate the degree days accumulated during a 150-day heating season if the average outside temperature is 17°F each day.

Seasonal Heating Degree Days = 150 days (65 – T_{out})

= 150 (65 – 17)

= **7,200 degree days**

In Illustration 7.2, we are assuming that the outside temperature remains at 17°F for all 150 heating days in a season. This is not realistic, but it explains the method to calculate the HDD. In a more realistic example we need to find the temperature difference for each day and add all the temperature differences (see Illustration 7.3).

Significance of HDD

Table 7.1 provides Seasonal HDDs for selected places in the United States. The higher HDD indicates a higher heat loss and, therefore, higher fuel requirements.

Illustration 7.3

Given the following set of average temperatures, by month, for State College, PA calculate the HDD for the heating season:

Month	Jan	Feb	Mar	Apr	May	Jun	Jul	Aug	Sep	Oct	Nov	Dec
Average outside temp.	25°F	28°F	37°F	48°F	59°F	67°F	71°F	70°F	62°F	51°F	41°F	31°F

Solution:

MONTH	TEMPERATURE DIFFERENCE (°F)	DAYS IN THE MONTH	DEGREE DAYS
January	65 − 25 = 40	31	1,240
February	65 − 28 = 37	28	1,036
March	65 − 37 = 28	31	868
April	65 − 48 = 17	30	510
May	65 − 59 = 6	31	186
June	65 − 67 = 0	(remember, if outside temperature is >65,	
July	65 − 71 = 0	HDD = 0)	
August	65 − 70 = 0		
September	65 − 62 = 3	30	90
October	65 − 51 = 14	31	434
November	65 − 41 = 24	30	720
December	65 − 31 = 34	31	1,054
Seasonal HDD			**6,138 degree days**

Table 7.1 *Annual Degree Days for Selected Places*

	DEGREE DAYS
Birmingham, AL	2,823
Anchorage, AK	10,470
Barrow, AK	19,893
Tucson, AR	1,578
Miami, FL	155
Pittsburgh, PA	5,829
State College, PA	6,345

SOURCE: *http://www.ncdc.noaa.gov/oa/climate/online/ccd/
nrmhdd.html*

Illustration 7.4

Mrs. Young is moving from Anchorage, Alaska (HDD = 10,780) to State College, PA (HDD = 6,000). Assuming the cost of energy per million BTUs is the same at both places, by what percentage will her heating costs change?

Solution: HDD in Anchorage, Alaska = 10,780
HDD in State College, PA = 6,000
Difference = 10,780 − 6,000 = 4,780

Savings in fuel costs are $\frac{4,780}{10,780} \times 100 = 44.3\%$

HDD is used to estimate the amount of energy required for residential space heating during a cool season, and the data are published in local newspapers or on the National Weather Service website.

R-Value

The R-value of thermal insulation depends on the type of material, its thickness, and its density. Generally, walls are not made up of just one material or one layer. R-values for most commonly used building materials are given in Table 7.2.

Table 7.2 shows that natural materials like stone and bricks are not good insulation materials, but most of the synthetic insulation products such as poly-

Table 7.2   *R-Values of Some Commonly Used Building Materials*

MATERIAL	R-VALUE $\left(\frac{ft^2\ °F\ h}{BTU}\right)$
Plywood, $\frac{3}{4}$ inch	0.94
Insulating sheathing, $\frac{3}{4}$ inch	2.06
Fiberglass, per inch (battens)	3.70
Polystyrene, per inch	5.00
Polyurethane board, per inch	7.00
Common brick, per inch	0.20
Gypsum board, $\frac{1}{2}$ inch (drywall or plasterboard)	0.45
Asphalt roof shingles	0.44
Stone, per inch	0.08
Plain glass, $\frac{1}{8}$ inch	0.03
Wood siding , $\frac{1}{2}$ inch	0.81
Cinder block, 12 inches	1.89

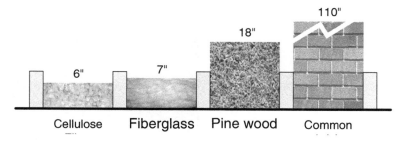

Figure 7.8

Thickness of insulation material required to obtain an R-value of 22.

styrene or polyurethane are very effective insulation materials. Higher thickness is required for materials with low R-value per inch to obtain a certain R-value for a wall. Figure 7.8 illustrates the thickness requirements of various insulation materials to obtain an R-Value of 22.

Types of Insulation

The right insulation material for a home depends upon the location of use and upon already existing material.

Fiberglass, made from molten glass spun into microfibers, is the most common type of insulation. It is usually pink or yellow and comes in the form of batts or rolled blankets (Figure 7.9).

Rock wool is literally made from rock—it is manufactured similar to fiberglass, but with molten rock instead of glass. The gray or brown fibers come in batts or blankets, or as shredded loose-fill.

Cellulose is made from recycled paper, such as newsprint or cardboard, shredded into small bits of fiber. It is treated with fire- and insect-resistant chemicals and is blown in as loose-fill (Figure 7.10).

Rigid foam insulation is applied directly to framing as rigid sheets. Several types of foam are available, some with post-consumer recycled content made from reclaimed fast-food containers and cups. Rigid foam is the insulation of choice where space is very limited, but a high R-value is needed. It can be installed on the interior or exterior of a wall, but on the inside it must be covered

Figure 7.9

A rolled blanket of fiberglass.

Figure 7.10

A photo of cellulose insulation being blown.
Source: *http://hem.dis.anl.gov/eehem/01/010909.html*

by a fire-resistant material like wallboard. Figure 7.11 shows a multilayer wall of a Penn State dorm room under construction. When the joints between panels are properly sealed, rigid foam insulation can act as both an air and vapor barrier.

One drawback to most foam insulation is that it deteriorates unless it is protected from prolonged exposure to sunlight and water. It also tends to be more expensive than most other types of insulation.

Synthetic insulation, usually polystyrene or polyurethane foam, is commonly used in rigid boards for insulating basements, cathedral ceilings, or sidewalls. Polyurethane foams are also high-performance insulating materials available as rigid boards as or sprayed-in-place systems.

Figure 7.11

A picture showing a multilayer insulation wall.

R-Value of a Composite Wall

Generally walls are made up of several layers of different materials. The R-value of a composite wall is calculated by adding the effective R-values of each of the layers of the wall. Figure 7.12 shows a wall made up of four layers—½-inch drywall inside for aesthetic purposes, real insulation in between the studs, ¾-inch plywood sheathing outside, and wood siding as the final exterior finish. The wall's composite R-value is calculated by adding the R-values of each layer. We need to add the R-values because all the walls are preventing heat loss together. R-values of each layer can be obtained from Table 7.2.

Plasterboard (½ inch)	=	0.45
Fiberglass 3.5 in. at 3.70/in.	=	12.95
Plywood (¾ inch)	=	0.94
Wood siding (½ inch)	=	0.81
Total R-Value of the Composite Wall	=	**15.15** $\frac{\text{ft}^2\,^\circ\text{F h}}{\text{BTU}}$

Illustration 7.5

A ceiling is insulated with 0.75-inch plywood, 2 inches of polystyrene board, and a 3-inch layer of fiberglass. What is the R-value for the ceiling?

Solution: The wall consists of three layers, and all three layers together prevent the heat loss. So we need to add the R-values of all three layers:

¾-inch plywood has an R-value of	0.94
2 inches of polystyrene at 5.0 per inch will have an R-value of	10.00
3 inches of fiberglass at 3.7 per inch will have an R-value of	12.10

So the R-value of the composite wall is 23.04 $\frac{\text{ft}^2\,^\circ\text{F h}}{\text{BTU}}$.

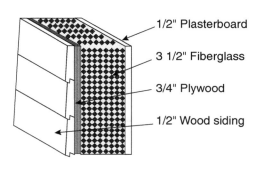

- 1/2" Plasterboard
- 3 1/2" Fiberglass
- 3/4" Plywood
- 1/2" Wood siding

Figure 7.13 shows the regions of the United States for insulation, and Figure 7.14 shows the U.S. Department of Energy Recommended Total R-Values for new construction houses by regions and by various parts of the house.

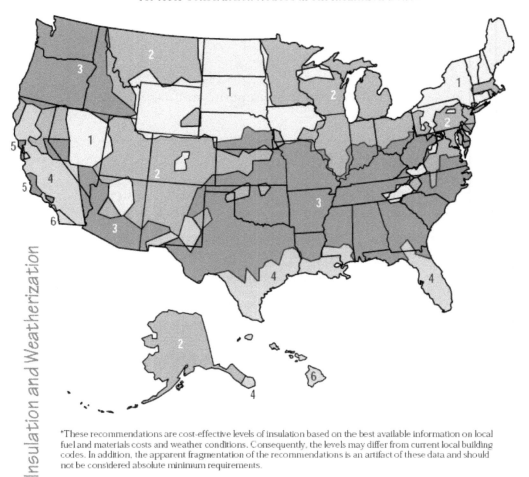

U.S. Department of Energy Recommended* Total R-Values
for New Construction Houses in Six Insulation Zones

*These recommendations are cost-effective levels of insulation based on the best available information on local fuel and materials costs and weather conditions. Consequently, the levels may differ from current local building codes. In addition, the apparent fragmentation of the recommendations is an artifact of these data and should not be considered absolute minimum requirements.

Insulation and Weatherization

Figure 7.13

The regions of the United States for insulation.

SOURCE: *http://www.eere.energy.gov/consumerinfo/energy_savers/pdfs/rvalue_map.pdf*

Zone	Gas	Heat pump	Fuel oil	Electric furnace	Ceiling		Wall(a)	Floor	Crawl space(b)	Slab edge	Basement	
					Attic	Cathederal					Interior	Exterior
1	✔	✔	✔		R-49	R-38	R-18	R-25	R-19	R-8	R-11	R-10
1				✔	R-49	R-38	R-28	R-25	R-19	R-8	R-19	R-15
2	✔	✔	✔		R-49	R-38	R-18	R-25	R-19	R-8	R-11	R-10
2				✔	R-49	R-38	R-22	R-25	R-19	R-8	R-19	R-15
3	✔	✔	✔	✔	R-49	R-38	R-18	R-25	R-19	R-8	R-11	R-10
4	✔	✔	✔		R-38	R-38	R-13	R-13	R-19	R-4	R-11	R-4
4				✔	R-49	R-38	R-18	R-25	R-19	R-8	R-11	R-10
5	✔				R-38	R-38	R-13	R-11	R-13	R-4	R-11	R-4
5		✔	✔		R-38	R-38	R-13	R-13	R-19	R-4	R-11	R-4
5				✔	R-49	R-38	R-18	R-25	R-19	R-8	R-11	R-10
6	✔				R-22	R-38	R-11	R-11	R-11	(c)	R-11	R-4
6		✔	✔		R-38	R-38	R-13	R-11	R-13	R-4	R-11	R-4
6				✔	R-49	R-38	R-18	R-25	R-19	R-8	R-11	R-10

(a) R-18, R-22, and R-28 exterior wall systems can be achieved by either cavity insulation or cavity insulation with insulating sheathing.
For 2 in. × 4 in. walls, use either 3½-in. thick R-15 or 3½-in. thick R-13 fiberglass insulation with insulating sheathing.
For 2 in. × 6 in. walls, use either 5½-in. thick R-21 or 6¼-in. thick R-19 fiberglass insulation.
(b) Insulate crawl space walls only if the crawl space is dry all year, the floor above is not insulated, and all ventilation to the crawl space is blocked. A vapor retarder (e.g., 4- or 6-mil polyethylene film) should be installed on the ground to reduce moisture migration into the crawl space.
(c) No slab edge insulation is recommended.
NOTE: For more information, see: Department of Energy Insulation Fact Sheet (D.O.E./CE-0180).
Energy Efficiency and Renewable Energy Clearinghouse, P.O. Box 3048, Merrifield, VA 22116; phone: (800) 363-3732; www.ornl.gov/roofs+walls/insulation/ins_11.html or contact Owens Corning, (800) GET-PINK (800-438-7465), www.owenscorning.com

Figure 7.14   *Department of Energy Recommended Total R-Values for New Construction Houses, by Regions and by Various Parts of the House.*
SOURCE: *http://www.eere.energy.gov/consumerinfo/energy_savers/pdfs/rvalue_map.pdf*

Calculation of Home Heat Loss

Heat loss from the surface of a wall can be calculated by using either Equation 7.1 or 7.3. Equation 7.1 gives the heat loss per hour, and Equation 7.3 gives the heat loss in full heating season. The heat loss from walls, windows, roof, and flooring should be calculated separately because of different R-values for each of these surfaces. If the R-values of walls and the roof are the same, the sum of the areas of the walls and the roof can be used with a single R-value.

Illustration 7.6

A house in Denver, CO has 580 ft² of windows (R = 1), 1,920 ft² of walls and 2,750 ft² of roof (R = 22). The walls are made up of wood siding (R = 0.81), 0.75-inch plywood, 3 inches of fiberglass insulation, 1.5 inches of polyurethane board, and 0.5-inch gypsum board. Calculate the heating requirement for the house for the heating season, given that the HDD for Denver is 6,100.

Solution: Heating requirement of the house = Heat loss from the house in the whole year. To calculate the heat loss from the whole house we need to calculate the heat loss from the walls, windows, and roof separately, and add all the heat losses.

Heat loss from the walls:

Area of the walls = 1,920 ft², HDD = 6,100, and the composite R-value of the wall needs to be calculated.

	R-value
Wood siding	0.81
¾-inch plywood	0.94
3 inches of fiberglass = 3 in. × 3.7/in.	10.70
1.5 inch of polyurethane board = 1.5 in. × 5/in.	7.50
½-inch plasterboard	0.45
Total R-value of the walls	20.40

$$Heat\,Loss\,from\,Walls = \frac{1,920\,ft^2 \times 6,100°F - days \times \frac{24\,h}{day}}{20.4\frac{ft^2\,°F\,h}{BTU}} = 13.92\,MMBTU$$

$$Heat\,Loss\,from\,Windows = \frac{580\,ft^2 \times 6,100°F - days \times \frac{24\,h}{day}}{1\frac{ft^2\,°F\,h}{BTU}} = 84.91\,MMBTU$$

$$Heat\,Loss\,from\,Roof = \frac{2,750\,ft^2 \times 6,100°F - days \times \frac{24\,h}{day}}{22\frac{ft^2\,°F\,h}{BTU}} = 18.3\,MMBTU$$

Total heat loss from the house = 13.92 + 84.91 + 18.30 = 116.53 MMBTU in a year or heating requirement is 116.83 million BTUs per year.

Fuel Choices for Home Heating

Various fuels such as natural gas, electricity, fuel oil, and so on, are used to heat a house. Figure 7.15 shows the choice of heating fuels of various households. More than 50 percent of the households in the United States use natural gas as their main heating fuel, and about 26 percent of the households use electricity to heat their homes. Each of these fuels is sold by different units. For example, natural gas is sold by cubic feet, oil by gallons, electricity by kWh, and coal by tons. The amount of heat a furnace can deliver is called its **capacity**, while the amount of energy actually used is called **consumption**. Heating units are manufactured and sold by their capacity; the monthly bills customers receive are for consumption. *(http://hem.dis.anl.gov/eehem/96/960309.html)*

Natural Gas

The heating capacity of gas heating appliances is measured in British Thermal Units per hour (BTU/h). (One BTU is equal to the amount of energy it takes to raise the temperature of one pound of water by 1° Fahrenheit.) Most heating appliances for home use have heating capacities of between 40,000 and 150,000 BTU/h. In the past, gas furnaces were often rated only on heat input; today the heat output is given.

Consumption of natural gas is measured in cubic feet (ft^3). This is the amount that the gas meter registers and the amount that the gas utility records when a reading is taken. One cubic foot of natural gas contains about 1,000 BTU of energy. Utilities often bill customers for CCF (100 cu. ft) or **therms** of gas used: One therm equals 100,000 BTUs. Some companies also use a unit of MCF, which is equal to 1,000 cu. ft. One MCF equals 1,000,000 BTUs (1 MMBTUs).

Fuel Oil

Several grades of fuel oil are produced by the petroleum industry, but only #2 fuel oil is commonly used for home heating. The heating (bonnet) capacity of oil heating appliances is the steady-state heat output of the furnace, measured in BTU/h. The typical oil-fired central heating appliances sold for home

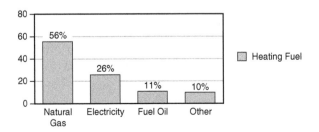

Figure 7.15

Choice of fuels by percent of households.

use today have heating capacities of between 56,000 and 150,000 BTU/h. Oil use is generally billed by the gallon. One gallon of #2 fuel oil contains about 140,000 BTU of potential chemical energy.

Electricity

The watt (W) is the basic unit of measurement of electric power. The heating capacity of electric systems is usually expressed in kilowatts (kW); 1 kW equals 1,000 W. A kilowatt-hour (kWh) is the amount of electrical energy supplied by 1 kW of power over a 1-hour period. Electric systems come in a wide range of capacities, generally from 10 kW to 50 kW. When converted to heat in an electric resistance heating element, 1 kWh produces 3,413 BTUs of heat.

Propane

Propane, or liquefied petroleum gas (LPG), can be used in many of the same types of equipment as natural gas. It is stored as a liquid in a tank at the house, so it can be used anywhere, even in areas where natural gas hookups are not available. Consumption of propane is usually measured in gallons; propane has an energy content of about 91.300 BTUs per gallon.

Each unit of fuel when burned gives different amounts of energy. The energy that is released when a unit amount of fuel is burned is called the **heating value.** The heating value of a fuel is determined under a standard set of conditions. A comparison of approximate heating values of various fuels is given in Table 7.3.

From Table 7.3, it should be noted that if a gallon of fuel oil is burned, one would get 140,000 BTUs. Similarly, a CCF of natural gas would fetch 100,000 BTUs. The important assumption here is that all the energy from the fuel is released and all the heat is available to heat the place. Generally when a fuel is burned in a furnace, not all the energy (heat) is available for the final end user. That is the energy efficiency of a furnace.

In other words, if the furnace efficiency is, say, 50 percent, we would have to put in twice as much fuel. The actual heating value of fuel oil is therefore 140,000 BTUs × 0.5 (efficiency) = 70,000 BTUs.

Table 7.3   *Heating Values of Commonly Used Heating Fuels*

FUEL	UNIT	HEATING VALUE (BTUs)
Natural gas	CCF (100 cu. ft) or Therm	100,000
Natural gas	MCF (1,000 cu. ft)	1,000,000
Fuel oil (#2)	Gallon	140,000
Electricity	kWh	3,412
Propane	Gallon	91,300
Bituminous coal	Ton	23,000,000
Anthracite	Ton	26,000,000
Hardwood	Cord	24,000,000

When a gallon of oil is burned, 70,000 BTUs of heat is actually available to the user. It can be clearly seen that as the efficiency of the furnace increases the amount of heat available increases proportionally. For example, if an oil furnace has an efficiency of 85 percent, the actual heat available from one gallon of oil would be 140,000 BTUs × 0.85 = 119,000 BTUs. The higher the efficiency, the less oil needs to be put into the furnace to get the same amount of heat output.

Most heating furnaces burn fuel and release hot combustion gases. The hot combustion gases heat the incoming cold air and go out through the chimney. In older furnaces, all the heat in the fuel is not released or not transferred to the cold air (or water, in the case of heat registers and water heaters) and therefore is lost through the chimney. The heated air or water distributes the heat throughout the house. Newer models of furnaces have gotten better at getting more of the heat into the cold air and, therefore, into your house.

Furnaces usually are not as efficient when they are first firing up as they are when running at steady-state—sort of like a car getting better mileage in steady highway driving than in stop-and-go city traffic (*http://homeenergysaver.lbl.gov/hes/aceee/afue.html*). What matters over the course of the year is the total useful heat the furnace delivers to your house versus the heat value of the fuel it consumes. This is kind of like measuring the gas mileage your car gets by asking how many miles you drove this year and dividing it by how many gallons of gas you bought. They call this measure for furnaces the AFUE (Annual Fuel Utilization Efficiency). The federal minimum-efficiency standards for furnaces and boilers took effect in 1992, requiring that new furnace units have an AFUE of at least 78 percent and new boiler units have an AFUE of at least 80 percent. In comparison, many old furnaces and boilers have AFUE ratings of only 55 to 65 percent. Table 7.4 gives the efficiencies of most efficient furnaces that are available in 2002–2003.

When buying a new furnace, make sure its heating capacity (output) is appropriate for your home. If the insulation and/or windows in your home have been upgraded since the old heating equipment was installed, you can probably use a much less powerful furnace or boiler. Oversized furnaces operate less efficiently because they cycle on and off more frequently; in addition, larger furnaces are more expensive to buy.

Table 7.4   *Efficiency Range of Some of the Most Efficient Furnaces in 2002–2003*

FUEL	FURNACE TYPE	EFFICIENCIES (%)
Natural gas	Hot air	93.0 – 96.6
Natural gas	Hot water	83.0 – 95.0
Natural gas	Steam	81.0 – 82.7
Fuel oil	Furnace	83.8 – 86.3
Fuel oil	Hot water	86.0 – 87.6
Fuel oil	Steam	82.5 – 86.0

Energy Costs

It is clear now that when a unit of fuel is burned not all of it is available to the end user, and that as the furnace efficiency increases higher amounts of heat will be available. An important question that needs to be addressed is how much it costs to buy the energy or heat to heat a place. Fuel is usually sold in gallons or CCF or kWh. Comparing the actual cost of energy to produce a certain amount of heat for the end user would be easy if the comparison were made on an energy basis rather than on a unit basis—that is, $/BTUs rather than $/gal or CCF or kWh.

Let's say we need one million BTUs to keep a place warm at a certain temperature. What would it cost to get those million BTUs from oil or gas or electricity? Let's assume that oil is sold for $1.25 a gallon. When a gallon of oil is burned the actual heat available to the end user is 140,000 BTUs × 0.85 (efficiency) = 119,000 BTUs. So, it costs $1.25 to get 119,000 BTUs or 0.119 MMBTUs. It costs $1.25/0.119 MMBTUs or $10.50/MMBTUs. We can write the above calculation into formula in Equation 7.4.

$$Actual\ Energy\ Cost = \frac{Fuel\ Cost\left(\dfrac{\$}{Unit\ of\ Fuel}\right)}{Heating\ Value\left(\dfrac{MMBTUs}{Unit\ of\ Fuel}\right) \times Efficiency} \qquad (7.4)$$

Electric resistance heat cost =

$$\frac{\dfrac{\$0.082}{\cancel{kWh}}}{\dfrac{0.003412\ MMBTUs}{\cancel{kWh}} \times 0.97\,(Efficiency)} = \$24.77\ per\ MMBTU$$

Natural gas (in central heating system) cost =

$$\frac{\dfrac{\$6.60}{\cancel{MCF}}}{\dfrac{1.0\ MMBTUs}{\cancel{MCF}} \times 0.90\,(Efficiency)} = \$7.33\ per\ MMBTU$$

Oil (in central heating system) cost =

$$\frac{\dfrac{\$1.25}{\cancel{Gal}}}{\dfrac{0.14\ MMBTUs}{\cancel{Gal}} \times 0.85\,(Efficiency)} = \$10.50\ per\ MMBTU$$

Annual Heating Cost

In Illustration 7.7, we see that the heat loss from the house (walls, windows, and the roof) was 116.53 MMBTUs. We also know that it costs $24.77 for 1 MMBTU if electrical resistance heating is used (see calculation in the previous section). The total cost for the heating can be calculated as follows:

$$Cost\ of\ Heating = (116.53\ MMBTUs) \times \frac{\$24.77}{MMBTUs} = \$2,886.44$$

The price of fuel oil is $10.50 per MMBTU. The annual heating cost would be:

$$Cost\ of\ Heating = (116.53\ MMBTUs) \times \frac{\$10.50}{MMBTUs} = \$1,223.57$$

Payback Period

Earlier sections illustrated that adding more insulation and improving the R-value of a wall would help in cutting heat loss. Less heat loss reduces the amount of fuel that needs to be burned, thereby reducing heating costs and protecting the environment. However, adding insulation often involves additional investment. The money invested can be recovered or paid back using the money saved because of the reduction of fuel usage. The time it takes to recover the additional cost through savings is called the payback period. A simple payback is the initial investment divided by annual savings after taxes. A simple calculation illustrates this term. If the R-value of the wall used in Illustration 7.6 is improved to R-23 by adding additional insulation, which costs $254, the heat loss can be reduced. The new heat loss after improvement can be calculated using Equation 7.3:

$$New\ Heat\ Loss\ from\ Walls = \frac{1,248\ \cancel{ft^2} \times 10,500°\cancel{F} - \cancel{days} \times \frac{24\ \cancel{h}}{\cancel{day}}}{23.0\frac{\cancel{ft^2}\ °\cancel{F}\ \cancel{h}}{BTU}} = \frac{13.7\ MMBTU}{Year}$$

Heat loss from the roof remains the same and is equal to 9.99 MMBTUs. Therefore, new annual total heat loss is only $13.7 + 9.99 = 23.69$ MMBTUs. The annual cost of heating after this improvement would be:

$$= \left(23.69\ \cancel{MMBTUs}\right) \times \frac{\$9.8}{\cancel{MMBTUs}} = \$232.16$$

The savings is $334.96 - $232.16 = $102.80 every year. Remember that to get this savings, an investment of $254 was made. So if this investment was paid off by the savings, it would take:

$$\frac{\$254.00}{\$102.80/year} = 2.47\ years$$

Illustration 7.7

A house in International Falls, MN (HDD = 10,500) consists of 1,248 ft^2 of walls with an R-value of 13 and 1,150 ft^2 of roof with an R-value of 29. The home is heated with natural gas. The AFUE is 0.90 and the price of natural gas is $0.88/CCF. What is the annual heating cost?

Energy cost per million BTUs from natural gas can be calculated using Equation 7.4:

$$Actual\,Energy\,Cost = \frac{Fuel\,Cost\left(\dfrac{\$}{Unit\,of\,Fuel}\right)}{Heating\,Value\left(\dfrac{MMBTUs}{Unit\,of\,Fuel}\right) \times Efficiency}$$

$$Actual\,Energy\,Cost = \frac{\dfrac{\$.88}{CCF}}{\dfrac{0.1\,MMBTUs}{CCF} \times 0.90(Efficiency)} = \$9.8\,per\,MMBTU$$

Heat required can be calculated from the heat loss. Heat loss from the house is from two sources: walls and the roof. Heat loss from each of these sources for a year (season) can be calculated by using Equation 7.3:

$$Heat\,Loss\,from\,Walls = \frac{1,248\,ft^2 \times 10,500°F - days \times \dfrac{24\,h}{day}}{13.0\dfrac{ft^2 \cdot °F\,h}{BTU}} = 24.19\,MMBTUs$$

$$Heat\,Loss\,from\,Roof = \frac{1,150\,ft^2 \times 10,500°F - days \times \dfrac{24\,h}{day}}{29\dfrac{ft^2 \cdot °F\,h}{BTU}} = 9.99\,MMBTUs$$

Total heat loss = sum of heat loss from the walls and the roof:

$$= 24.19 + 9.99 = 34.18\,MMBTUs$$

Annual heating cost = Annual heat loss (MMBTUs) × Actual energy cost ($/MMBTU):

$$= \left(34.18\,MMBTUs\right) \times \frac{\$9.8}{MMBTUs} = \$334.96$$

The payback period is 2.47 years. Shorter payback periods indicate that the additional investment can be paid off quickly and the homeowner can start saving money after that.

The formula below will help you to estimate the cost effectiveness of adding insulation in terms of the "years to payback" for savings in heating costs.

$$Years\ to\ Payback = \frac{C_i \times R_1 \times R_2 \times E}{C_e \times [R_2 - R_1] \times HDD \times 24} \qquad (7.5)$$

Where:

C_i = **Cost of insulation in $/square feet.** Collect insulation cost information; include labor, equipment, and vapor barrier if needed.

C_e = Cost of energy, **expressed in $/BTUs.** To calculate this, divide the actual price you pay per gallon of oil, kilowatt-hour (kWh) of electricity, gallon of propane, or therm (or per one hundred cubic feet [CCF]) of natural gas by the BTU content per unit of fuel.

E = Efficiency of the heating system. For gas, propane, and fuel oil systems, this is the Annual Fuel Utilization Efficiency, or AFUE.

R_1 = Initial R-value of section

R_2 = Final R-value of section

$R_2 - R_1$ = R-value of additional insulation being considered

HDD = Heating degree days/year. This information can usually be obtained from your local weather station, utility, or oil dealer.

24 = Multiplier used to convert HDD to heating hours (24 hours/day).

Equation 7.5 works only for uniform sections of the home. For example, you can estimate years to payback for a wall or several walls that have the same R-values, if you add the same amount of insulation everywhere. Ceilings, walls, or sections of walls with different R-values must be figured separately.

Sources

http://www.aceee.org/consumerguide/topfurn.htm

http://www.eere.energy.gov/consumerinfo/factsheets/ea3.html

Elements of an energy efficient house, Factsheet, DOE/GO-102-1070 PS-207, July 2000, pp. 1–8.

http://www.eere.energy.gov/consumerinfo/factsheets/ed3.html

http://www.eere.energy.gov/consumerinfo/factsheets/eb9.html

http://www.eere.energy.gov/consumerinfo/factsheets/ec1.html

Insulation, Factsheet, Energy Efficiency and Renewable Energy, U.S. Department of Energy DOE/CE-1080, 2002.

Illustration 7.8

Mr. Energy Conscience (who lives in East Lansing, MI with an HDD of 7,164) wants to know how many years it will take to recover the cost of installing additional insulation in his attic. He renovated his attic and increased the level of insulation from R-19 to R-30 by adding additional insulation. He has a gas furnace with an AFUE of 0.88 and pays $0.95/CCF for natural gas. The attic insulation costs $340 to cover 1,100 sq. ft.

The payback period is given by the Equation 7.5:

$$Years\ to\ Payback = \frac{C_i \times R_1 \times R_2 \times E}{C_e \times [R_2 - R_1] \times HDD \times 24}$$

$R_1 = 19$; $R_2 = 30$, and $R_2 - R_1 = 30 - 19 = 11$

$HDD = 7,164$ and $E = 0.88$

The most important part of this problem is to determine the cost of insulation per 1 sq. ft (C_i) and cost of energy per one BTU:

$$C_i = \frac{\$340}{1,100\ sq.\ ft} = \$0.31\ per\ sq.\ ft$$

$$C_e = \frac{\$0.95}{1\,CCF \times \frac{100,000\ BTUs}{1\,CCF}} = \$0.0000095\ per\ BTU$$

NOTE: The cost for 1 BTU is a very small number.

Substituting the values in the equation,

$$Years\ to\ Payback = \frac{0.31 \times 19 \times 30 \times 0.88}{0.0000095 \times [11] \times 7,164 \times 24} = 8.65\ years$$

After 8.65 years, Mr. Energy Conscience can start saving money for himself for the rest of the period that he lives in that home. During the entire period the energy that he is not using can help the environment.

http://www.naima.org/pages/resources/library/pdf/BI403.PDF
http://www.naima.org/pages/resources/library/html/BI409.HTML
http://www.eere.energy.gov/consumerinfo/energy_savers/insulation.html
http://www.aceee.org/consumerguide/topfurn.htm
http://www.eere.energy.gov/consumerinfo/energy_savers/pdfs/energysavers_34430.pdf

questions

1. Define conduction, convection, and radiation. Give two examples of each of these mechanisms as applied to home heating.

2. Define R-value and explain the factors that determine the R-value of a material.

3. List any three common insulating materials.

4. Explain the significance of Heating Degree Days (HHD). ·

5. What are typical prices ($/MMBTUs) of electricity, natural gas, and #2 fuel oil used for home heating in your region?

6. What is the average number of Heating Degree Days and cooling degree days in your location?

7. A friend of yours is planning on adding more insulation to his home. Since his furnace efficiency is low, does it make sense to do this? And what impact does this have on the payback period?

multiple choice questions

1. Heat is lost through walls, windows, and roof of a house. The primary mechanism by which the heat is lost through the walls is
   a. Convection
   b. Radiation
   c. Conduction
   d. None of the above

2. This method of heat transfer involves actual movement of molecules.
   a. Conduction
   b. Convection
   c. Radiation

3. The difference in temperature across a wall causes heat loss to occur by
   a. Conduction
   b. Convection
   c. Radiation
   d. None of the above

4. The R-value of a thermal insulation material depends on its
   a. Density
   b. Thickness
   c. Type
   d. All the above

5. The bulk of home energy costs is due to
   a. Heating, ventilation, and air conditioning (HVAC)
   b. Refrigerator
   c. Water heating
   d. Lighting

6. Of the following places, where does the most home heat loss occur in a typical home?
   a. Windows
   b. Doors
   c. Through cracks in walls, windows, and doors
   d. Basement
   e. Roof

7. How many degree days are accumulated in a seven-day period where the average temperature outside is 35 degrees Fahrenheit?
   a. 245
   b. 210
   c. 249
   d. 6,000

8. _____ is an index of fuel consumption indicating how many degrees the mean temperature fell below 65 degrees for the day.
   a. CCD
   b. HDD
   c. R-value
   d. COP

9. Windy air picking up heat from a glass window is an example of heat transfer by
   a. Conduction
   b. Convection
   c. Radiation
   d. None of the above

10. In a location where the cost of energy is less, the payback period for adding more insulation would be
    a. Higher (longer)
    b. Lower (shorter)

11. Birmingham, AL has an HDD of 3,000 and State College, PA has an HDD of 6,000. The payback period for adding more insulation would be lower in
    a. State College, PA
    b. Birmingham, AL

12. Oil is sold in a particular region at $1.40 per gallon and natural gas at $8.47 per MCF. Which fuel is cheaper to use?
    a. Oil
    b. Natural gas

13. In Reading, PA, oil is sold at $1.40 per gallon and natural gas at $0.975 per CCF. Assuming the efficiency of oil furnace is 80% and that of a natural gas furnace is 90%, determine the savings per million BTUs using the actual energy costs.
    a. $3.08
    b. $0.19
    c. $1.67
    d. $2.80

14. The heating fuel used by most households in the U.S. is
    a. Natural gas
    b. Fuel oil
    c. Electricity
    d. Coal

15. If Jon wants to reduce his heating cost then he should
    a. Increase the insulation
    b. Buy a small house
    c. Move out to a place where HDD is low
    d. Any of the above

16. Increasing the insulation after a certain level will not bring proportional savings in fuel costs.
    a. True
    b. False

problems for practice

Heating Degree Days

(Assume HDD = 6,000 degree days and the inside temperature is 65°F unless otherwise stated.)

1. Calculate the heating degree days (HDD) for one day when the average temperature outside is 13°F.

2. Calculate the HDD for one day when the average temperature outside is 2°C.

3. Given the following data, calculate the HDD for the week:

Day	Average Temperature
Sunday	49°F
Monday	47°F
Tuesday	51°F
Wednesday	60°F
Thursday	65°F
Friday	67°F
Saturday	58°F

4. For the month of January in State College, PA, the average daily temperature was 25°F. Calculate the HDD for the month of January.

5. Given the following set of average temperatures by month for State College, PA, calculate the HDD for the heating season:

Jan.	Feb.	Mar.	Apr.	May	Jun.	Jul.	Aug.	Sep.	Oct.	Nov.	Dec.
25°F	28°F	37°F	48°F	59°F	67°F	71°F	70°F	62°F	51°F	41°F	31°F

6. If Ms. S. Belle moves from Birmingham, AL (HDD = 2,800) to State College, PA, (HDD = 6,000), how much can she expect her heating bill to increase?

Composite R-Values

If R-values are given in the problem statement, use those. If R-values are not given, use those from Table 7.3.

7. A wall consists of 0.5-inch wood siding (R = 0.81), 0.75-inch plywood (R = 0.94), 3.5 inches of fiberglass (R = 13.0), and 0.5-inch plasterboard (R = 0.45). What is the composite R-value of the wall?

8. A basement wall is made up of 3 inches of brick (R = 0.20 per inch), a layer of cinder block (R = 1.89), and a layer of 0.5-inch drywall (R = 0.45). What is the composite R-value for the basement wall?

9. What is the R-value of a wall that is made up of wood siding (R = 0.81), 5 inches of fiberglass (R = 3.70 per inch), and a layer of 0.5-in. drywall (R = 0.45)?

10. A ceiling is insulated with 0.75-in. plywood, 2 inches of polystyrene board, and a 3-in. layer of fiberglass. What is the R-value for the ceiling?

11. A wall is made up of 8 inches of stone, 3 inches of polyurethane board, and 0.75 inch of plywood. Calculate the composite R-value for the wall.

Heat Loss

12. What is the area of a wall that measures 20 ft long by 8 ft high?

13. The floor dimensions of a room are 16 ft by 12 ft. The ceiling is 8 ft high. What is the wall area for this particular room? (Do not worry about the floor or ceiling.)

14. A room is 20 ft by 12 ft with 10 ft ceilings. There are two windows in the room, as well as a door. One window measures 3 ft wide by 4 ft high. The other is 6 ft wide by 4 ft high. The door is 4 ft wide and 7 ft high. Calculate the areas of the door, windows, and walls. (Do not worry about the floor or ceiling.)

15. Calculate the heat loss per hour for a 10 ft by 8 ft wall, insulated to R-value 22. The inside temperature is maintained at 70°F. The temperature outside is 43°F.

16. What is the heat loss per hour through a wall measuring 12 ft by 9 ft insulated to R-19? The inside is maintained at 72°F; the outside is 27°F.

17. A room with a total area of 360 ft² is insulated to R-22. The furnace is set at 65°F. How many BTUs are lost through the room in 12 hours when the average outside temperature is 37°F?

18. The thermostat in a small house is set at 68°F. The area covered by the windows (R-1) is 120 ft², and the area covered by walls (R-19) is 776 ft². The roof (R-22) area is 900 ft². Calculate the heat loss per hour through the house when the outside temperature is 33°F.

19. For a 36-hour period, the average temperature outside was –4°F. How much heat is lost through a 246 ft² wall insulated to R-19 if the inside temperature is kept at 70°F?

20. The average heating requirement for State College, PA, is 6,000 degree days (°F day). How much heat is lost through a wall (R = 19) with an area of 184 ft² throughout the entire heating season?

21. Calculate the heat lost through a 2 ft by 3 ft window (R = 1) in Anchorage, AK, if the heating requirement was 11,000 degree days.

22. For the 150-day heating season in Roanoke, VA, the average temperature was 47°F. How much heat is lost through a 176 ft² wall (R = 16) during the entire season?

23. In Fargo, ND, the heating season lasts about 220 days, and the average outside temperature is around 27°F. How much heat is lost through an 8 ft by 6 ft window (R = 1) during the heating season?

24. A 160 ft² wall in State College, PA consists of a 3-inch brick layer, a layer of 12-inch cinder block, 2 inches of fiberglass insulation, and 0.5-inch drywall (gypsum board). How much heat is lost through the wall during the heating season?

25. A house in State College, PA has 580 ft² of windows (R = 1), 1,920 ft² of walls, and 2,750 ft² of roof (R = 22). The walls are made up of wood siding (R = 0.81), 0.75-inch plywood, 3 inches of fiberglass insulation, 1.5-inch polyurethane board, and 0.5-inch gypsum board. Calculate the heating requirement for the house for the heating season.

26. A single-story house in Anchorage, AK (HDD = 11,000) is 50 ft by 70 ft with an 8 ft high ceiling. There are six windows (R = 1) of identical size, 4 ft wide by 6 ft high. The roof is insulated to R-30. The walls consist of a layer of wood siding (R = 0.81), 2 inches of polyurethane board (R = 6.25 per inch), 4 inches of fiberglass (R = 3.70 per inch), and a layer of drywall (R = 0.45). Calculate the heat loss through the house (not counting the floor) for the season. (A good estimate for the area of the roof is 1.1 times the area of the flat roof.)

Heating and Fuels

Assume the furnace efficiency is 100%, unless otherwise noted.

27. How many BTUs are provided by 34 CCF of natural gas?

28. How many BTUs are provided by 83 gallons of heating oil?

29. The seasonal heating requirement for a house is 236 MMBTUs. How much natural gas is required to heat the home?

30. How many gallons of oil must be purchased to meet the heating needs of a house requiring 173 MMBTUs?

31. 106 CCF of natural gas is equivalent to how many kWh of electricity?

32. A house in State College, PA has 580 ft^2 of windows (R = 1), 1,920 ft^2 of walls (R = 19), and 2,750 ft^2 of roof (R = 22). Calculate how much natural gas is required to heat this house for the heating season if the furnace efficiency is 90%.

33. A house in Bismark, ND (HDD = 9,000) has 860 ft^2 of windows (R = 1), 2,920 ft^2 of walls (R = 19), and 3,850 ft^2 of roof (R = 22). Calculate how much heating oil is required to heat this house for the heating season. The furnace efficiency is 80%.

Economics

34. If the market price of natural gas is $11.11 per MCF, what is the cost per MMBTU?

35. Heating oil is currently selling for $1.08/gal. What is the cost per MMBTU?

36. The price of electricity is $0.12/kWh in your region. Calculate the cost per MMBTU.

37. A ton of low-sulfur, Western coal from the Powder River Basin sells for $6.50. Calculate the cost per MMBTU.

38. Natural gas is selling for $1.24/CCF. If your furnace efficiency is 82%, what is your real cost of natural gas per MMBTU?

39. Your old oil furnace runs at about 68% efficiency. If you buy your oil for $1.02/gal, calculate your actual cost on an MMBTU basis.

40. Natural gas costs $9.74/MCF. Heating oil costs $0.99/gal. The natural gas furnace runs at 90% efficiency, and the oil furnace runs at 80% efficiency. Which fuel is cheaper?

41. If the King of Town installs new insulation in his castle for $6,500, what is the payback period if, by installing this insulation, he estimates he will save $800 in fuel costs every year?

42. Marzipan estimates she can save $125/year by adding additional insulation to her roof, which costs $2,400. Is this a good investment?

43. For a house in Hackensack, NJ (HDD = 4,600), the installed cost to upgrade from R-13 to R-22 is $0.60/ft^2. The AFUE for the oil furnace is 0.78, and heating oil costs $1.13/gal. How long will it take to recover the initial investment?

44. Lt. Dave Rajakovich has 1,200 ft^2 of roof in his home in Pittsburgh, PA (HDD = 6,000). He is considering upgrading the insulation from R-16 to R-22. The estimate from the contractor was $775. His home is heated with natural gas. Last year, the average price he paid for natural gas was $9.86/MCF. Assuming an AFUE of 86%, how long will it take Dave to recover his investment?

45. A house in Chicago, IL (HDD = 6,500) consists of the following:

- 1,866 ft^2 of walls with an R-value of 19
- 600 ft^2 of stone walls, 13″ thick (R = 0.08 per inch)
- 1,460 ft^2 of roof with an R-value of 22
- 626 ft^2 of single-pane windows (R = 1)

If new, energy-efficient windows (R = 6) are installed for $8,000, how long will it take to recover the initial investment? The home is heated with heating oil and pays $1.30/gal. The furnace efficiency is 80%.

46. A house in International Falls, MN (HDD = 10,500) consists of the following:

- 1,248 ft^2 of walls with an R-value of 22
- 860 ft^2 of stone walls, 13″ thick (R = 0.08 per inch)
- 1,340 ft^2 of roof with an R-value of 26
- 570 ft^2 of single-pane windows (R = 1)

The owner would like to add insulation to the stone walls to make the new R-value 22. She gets an estimate for $0.75/ft^2. Her home is heated with natural gas. The AFUE is 0.90, and the price of natural gas is $0.88/CCF. Is this a good investment?

Home Heating Basics Puzzle

Across

1. Involves actual movement of molecules to carry the heat

4. An indicator of heating requirement at a location. The higher this value at a place, the less time it takes to recover the additional insulation costs

7. A material used in sealing cracks in windows, doors and walls

8. The higher the difference in this, the higher is the heat loss from a house

10. A material that has higher conductivity

11. These materials generally do have higher thermal conductivity

13. A material that has good insulation properties and is loosely filled

14. A place in the house that can be easily insulated after construction

Down

2. The famous equation to calculate heat loss through a wall accounts for this type of heat loss

3. The method of heat transfer by electromagnetic waves

4. This device can provide both heating and cooling

5. Feels like powdered glass packed but is an insulating material

6. A house built with these is not very energy efficient

9. The higher this value, the lower is the heat loss through a wall

12. The higher this value, the higher is the heat loss from a house

Home Heating Systems

goals

- To gain basic understanding of operating principles of various types of heating systems

- To know the main advantages and disadvantages of various heating systems

- To understand the energy efficiencies of each of these heating systems

- To know ways to improve the energy efficiency of the heating systems

ore than 90 million single-family, multi-family, and mobile-home households encompass the residential sector. Households use energy to cool and heat their homes, to heat water, and to operate many appliances such as refrigerators, stoves, televisions, and hot tubs. Figure 8.1 shows how U.S. households use energy. Nearly half of the energy used at home is for space heating. Conventional space heating systems in the United States are responsible for a billion tons of carbon dioxide (CO_2) and about 12 percent of the sulfur dioxide and nitrogen oxides emitted by the nation. Reducing the use of conventional energy sources for heating is the single most effective way to reduce your home's contribution to global environmental problems.

The energy sources utilized by the residential sector include electricity, natural gas, fuel oil, kerosene, liquefied petroleum gas (propane), coal, wood, and other renewable sources such as solar energy. The primary fuel that U.S. households use is shown in Figure 8.2.

Home heating systems are classified based on the fuel and/or the method by which the heat is transferred and distributed into the house. Furnaces, boilers, or electric resistance heat powered by conventional fuels supply most of the heating in homes and commercial buildings today. To a lesser (but increasing) extent, heat pumps are also used to provide space heating.

A central heating system has a primary heating appliance such as a furnace or a boiler located in an out-of-the-way spot such as a basement or garage. It delivers heat throughout the house, either by pumping warmed air through a system of air ducts or sending hot water or steam through pipes to room radiators or convectors. With both forced-air and gravity systems, one or more thermostats turn the heating (or cooling) plant off and on, operated either manually or automatically as room temperatures rise and fall. Homes without central

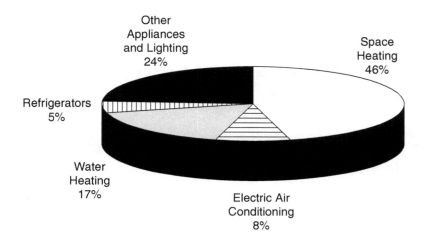

Figure 8.1   *Average U.S. household energy consumption.*

DATA SOURCE: *ftp://ftp.eia.doe.gov/pub/consumption/residential/2001ce_tables/enduse_consump.pdf*

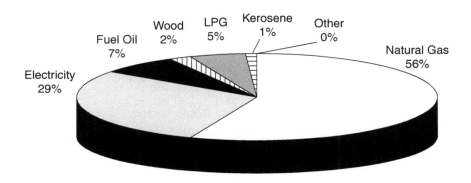

Figure 8.2

Main residential heating fuel of U.S. households.

DATA SOURCE: *ftp.eia.doe.gov/pub/consumption/residential/2001ce_tables/enduse_consump.pdf*

heating normally utilize electric baseboard heaters or, in some cases, in-the-wall or in-floor gas heaters or radiant heat.

Central Ducted Air Systems

Ducted air systems are the most common type of central heating and cooling used. If a home has a central air conditioner, heat pump, or furnace, generally it is a ducted air system. A duct can be considered as a rectangular pipe. There are two main types: forced-air and gravity.

Forced Air Heating Systems

Almost 35 million homes in America are heated by natural gas-fired, forced-air heating systems, by far the most popular form of central heating (Figure 8.3). Unfortunately, nearly all of these households have been sending 30 percent or more of their energy dollars up the furnace flue, contributing to an additional 50 tons of carbon dioxide every year per household. Most conventional forced-air furnaces operate at very low efficiencies of about 50 percent. This is like utilizing only 50 percent of the energy that we buy and feed into the furnace.

In a forced-air heating system, room air (cooler) is drawn by a fan or a blower through return air registers and ductwork and passes through a filter (to remove any dust particles) into a furnace, where the air is heated (Figure 8.4). The warmed air is then blown back to the rooms through a system of supply ducts and registers. With a forced-air system, a furnace warms air, an air conditioner cools air, or a heat pump either warms or cools air, then a blower forces the air through the system. Therefore, the same duct system can be used for both heating and cooling.

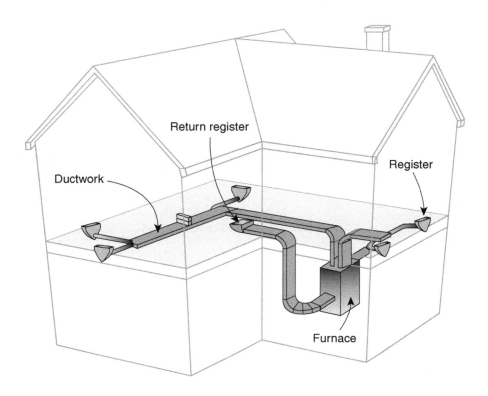

Figure 8.3 *Ducted heat distribution system.*

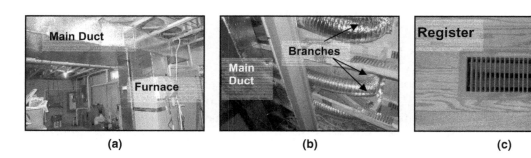

(a)   (b)   (c)

Figure 8.4 *Illustration of forced-air duct heat distribution system:*
(a) main duct, (b) branches, and (c) registers.

    With a gravity furnace, convection currents (caused by the natural tendency of heated air to rise) carry heated air through the system from a furnace that is located on or below the main floor level. Gravity systems, which are somewhat older, do not have blowers and tend to have very large air ducts; they can only deliver warmed air.

Most furnaces are gas-fired, but other fuels include oil, coal, wood, and electricity. With a conventional furnace, natural gas is piped to a burner located inside a combustion chamber. There the gas is mixed with air, then ignited by a pilot light, a spark, or a similar device controlled by a thermostat. The flame produces high temperature gases that exchange heat with cold incoming air in a device called a "heat exchanger." Exhaust gases given off by burners are vented to the outside through a flue that goes up through the roof or, with newer high-efficiency models, out through a wall. An electric forced-air furnace uses heating elements rather than burners to heat the air in the heat exchanger.

Heat Distribution

In a hot-air system, warm air is distributed via a main duct (Figure 8.4a) and a series of branches that lead to individual rooms or zones. Where the branches (Figure 8.4b) meet the main duct, heat is controlled by dampers (act as valves for air flow), which open or close to release or block hot air from entry. These dampers, usually motorized, are run by thermostatic controls at each zone. Individual registers (Figure 8.4c) may also be closed to block heat, but this is a less efficient use of the energy and heat produced than using thermostatic or automatic controls.

An upward-flow furnace draws cold air in through the bottom and sends heated air out the top. Upward-flow furnaces are often used in houses that have basements or that deliver heat through overhead ductwork. Downward-flow or counter-flow furnaces draw cool return air through the top and deliver heated air out the bottom. This type is favored where there is no basement or where air ducts are located in the floor.

A forced-air heating system can be combined with air conditioning (for cooling), a humidifier (for maintaining proper moisture balance), and an air filter (for purifying the air). Ductwork is generally metal, wrapped with insulation to help keep heat in. In some cases, flexible insulation-style ductwork is preferred. This system has several advantages and disadvantages (Table 8.1).

Radiant Heating Systems

Baseboard Radiators

In baseboard hydronic heating systems (shown in Figure 8.5) water is heated in a gas-fired or oil-fired furnace located in the basement. The heated water is distributed through pipes into baseboards in various rooms. The heat is then delivered through radiation and convection. Although these are called radiant heating systems, most of the heat delivered is by convection. Heat delivery into rooms or zones can be controlled by flaps or louvers.

Table 8.1	Advantages and Disadvantages of Forced-Air Duct System
ADVANTAGES	**DISADVANTAGES**
■ Air ducts and registers distribute heated air from a central furnace, providing rapid heat delivery. ■ The system can also be used to filter and humidify the household air and to provide central air conditioning. ■ The system circulates air for ventilation.	■ Air coming from the heating registers sometimes feels cool (especially with certain heat pumps), even when it is warmer than the room temperature. ■ There can also be short bursts of very hot air, especially with oversized units. ■ Ductwork may transmit furnace noise and can circulate dust and odors throughout the house. ■ Ducts are also notoriously leaky, typically raising a home's heating costs by 20% to 30%.

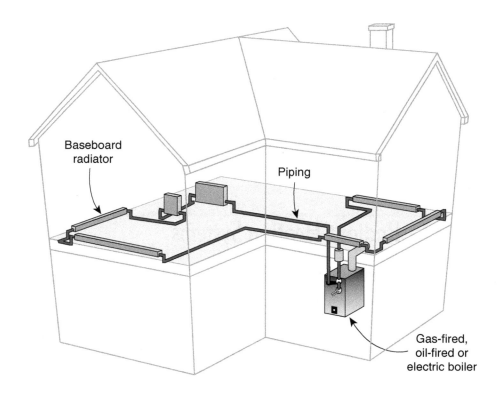

Baseboard radiator

Piping

Gas-fired, oil-fired or electric boiler

Figure 8.5

A baseboard radiator.

196   *chapter* 8   *Home Heating Systems*

A picture of a baseboard heater is shown in Figure 8.6. Closely spaced metallic sheets called "fins" increase the surface area for efficient heat transfer into the room. Some advantages and disadvantages are listed in Table 8.2.

Radiant Floor Heat

There are three types of radiant floor heat: radiant air floors (air is the heat-carrying medium); electric radiant floors; and hot water (hydronic) radiant floors. All three types can be further subdivided by the type of installation: those that make use of the large thermal mass of a concrete slab floor or lightweight concrete over a wooden subfloor ("wet installations") and those in which the installer "sandwiches" the radiant floor tubing between two layers of plywood or attaches the tubing under the finished floor or subfloor ("dry installations").

Because air cannot hold large amounts of heat, radiant air floors are not cost effective in residential applications and are seldom installed.

Figure 8.6

Baseboard heater.

Table 8.2	Advantages and Disadvantages of Radiant Baseboard Heating Systems

ADVANTAGES	DISADVANTAGES
■ Generally, it operates quietly. ■ It delivers constant heat and doesn't stir up allergens or dust. ■ Because it warms people and objects rather than just air, it feels warm even if a door is opened or a room is somewhat drafty or slightly cooler than normal. ■ There is less heat loss (waste) compared to the forced air system because there is no leakage.	■ Cannot be used for cooling ■ High installation costs ■ Interference with furniture placement ■ Air entrapment can reduce efficiency

Electric radiant floors are usually only cost effective if your electric utility company offers time-of-use rates. Time-of-use rates allow you to "charge" the concrete floor with heat during off-peak hours (approximately 9 P.M. to 6 A.M.). If the floor's thermal mass is large enough, the heat stored in it will keep the house comfortable for eight to ten hours without any further electrical input. This practice saves a considerable number of energy dollars compared to heating at peak electric rates during the day.

Hydronic (liquid) systems, popular and cost-effective systems for heating-dominated climates, have been in extensive use in Europe for decades (Figure 8.7). Hydronic radiant floor systems pump heated water from a boiler through tubing laid in a pattern underneath the floor. The temperature in each room is controlled by regulating the flow of hot water through each tubing loop via a system of zoning valves or pumps and thermostats. Wet installations are the oldest form of modern radiant floor systems. In a wet installation, the tubing is embedded in the concrete foundation slab, or in a lightweight concrete slab on top of a subfloor, or over a previously poured slab.

However, due to recent innovations in floor technology, dry floors have been gaining a lot of popularity over wet floors, to a great degree because a dry floor is faster and less expensive to build. There are several ways to make a dry radiant

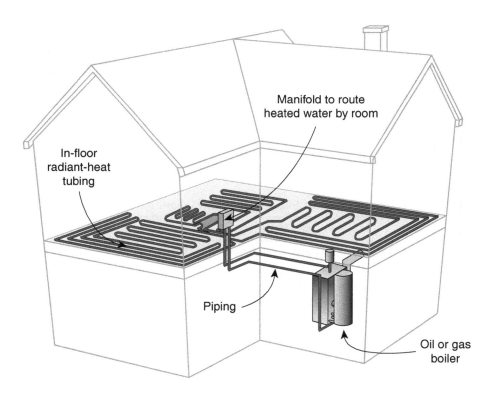

Figure 8.7   *A hydronic floor heating system.*

floor. Some dry installations involve suspending the tubing underneath the sub-floor between the joists. This method usually requires drilling through the floor joists in order to install the tubing. Reflective insulation must also be installed under the tubes to direct the heat upward. Tubing may also be installed from above the floor, between two layers of subfloor.

A new generation of in-floor hydronic heating that employs corrosion-proof, hot-water tubing has enjoyed widespread popularity in recent years. With this type of system, heat is evenly distributed and floors are warm. A variety of heating equipment may heat water: natural gas or propane water heater or boiler, electric boiler, wood boiler, heat pump, solar collector, or even geothermal energy.

Tubing for a hydronic system may be installed in a conventional concrete slab or in a lightweight, gypsum-cement slab. It can also be stapled to the undersides of subflooring as shown in Figure 8.8.

Floor Coverings

Although ceramic tile is the most common floor covering for radiant floor heating, almost any floor covering can be used. However, some perform better than others. Common floor coverings like vinyl and linoleum sheet goods, carpeting, wood, or bare concrete are often specified.

Figure 8.8 *Layout of the tubing in a hydronic floor system (subfloor).*

However, it is wise to always remember that anything that can insulate the floor also reduces or slows the heat entering the space from the floor system, which in turn increases fuel consumption. If carpeting is required, a thin carpet with dense padding is preferred. If some rooms, but not all, will have a floor covering, then those rooms should have a separate tubing loop to make the system heat these spaces more efficiently. That is because the water flowing under the covered floor will need to be hotter to compensate for the floor covering.

Most radiant floor references also recommend using laminated wood flooring instead of solid wood, thus reducing the possibility of the wood shrinking and cracking from the drying effects of the heat. While solid wood flooring can be used, the installer is strongly advised to be very familiar with radiant floor systems before attempting to install natural wood flooring over a radiant floor system. Most manufacturers and manuals relating to radiant floors offer guidelines to help you resolve these issues.

Types of Tubing

Older radiant floor systems used either copper or steel tubing embedded in the concrete floors. Unless the builder coated the tubing with a protective compound, a chemical reaction between the metal and the concrete often led to corrosion of the tubing and to eventual leaks. Major manufacturers of hydronic radiant floor systems now use cross-linked polyethylene (PEX) or rubber tubing with an oxygen diffusion barrier. These materials have proven more reliable than the older choices in tubing. Fluid additives also help protect the system from corrosion.

There have been recent reports of problems with rubber tubing produced by one chemical manufacturer. Leaks develop at the metal connections or fittings, and in some cases the tubing becomes rigid and brittle. It is still not clear what causes this problem, but theoretically excessively high water temperatures may be to blame. Tightening the connections and clamps only temporarily fixes the leaks. Remember, this problem only concerns a specific brand of rubber tubing; it does not have anything to do with the PEX tubing, which has performed very reliably for many decades. Since the price of copper tubing is considerably lower now than several years ago, it is again gaining some popularity because of its superior heat transfer abilities over plastic-based tubing.

Controlling the System

A radiant floor that uses a concrete slab takes many hours to heat if it is allowed to become cold. It can be very inconvenient waiting for the slab to heat up so it can heat the space. Consequently, most radiant floor systems are not permitted to go into a very deep night setback. Depending on how the floor is constructed, the time it takes to reheat the floor is sometimes longer than the occupant's sleep period.

ADVANTAGES	DISADVANTAGES
■ Radiant floor systems allow even heating throughout the whole floor, not just in localized spots as with wood stoves, hot air systems, and other types of radiators. ■ The room heats from the bottom up, warming the feet and body first. ■ Radiant floor heating also eliminates the draft, dust, and allergen problems associated with forced-air heating systems. ■ With radiant floor heating, you may be able to set the thermostat several degrees lower, relative to other types of central heating systems. ■ There are no heat registers or radiators to obstruct furniture arrangements and interior design plans.	■ Does not respond quickly to temperature settings ■ Relatively expensive to install but can save money in the long run ■ Requires professional installers

Many floor systems are also controlled by a floor thermostat instead of a wall thermostat. The system is also often designed to keep the circulation pump(s) running while the thermostat only controls the boiler's burner. Other more sophisticated, types of controls sense the floor temperature, outdoor temperature, and room temperature to keep the home comfortable. Such a system may also use less fuel.

Although radiant floor systems are usually heated by a boiler, they can also be heated with a geothermal heat pump. Such a system offers even greater energy savings in climates where the heating and cooling loads are similar in size. Another alternative for small houses, or those with small heating loads, is to use an ordinary gas water heater to supply the radiant floor system. Table 8.3 lists some of the advantages and disadvantages of radiant floor heating systems.

Direct or In-Situ Heating Systems

Instead of generating heated air or water at a central location and then distributing it throughout the home, some systems generate heat where it is needed locally. The most common method is electric baseboard heat. Other ways include kerosene heat, wood-burning stoves, and fireplaces burning wood, coal, or natural gas. These systems can heat the whole house, part of the house, or a single room.

Electric Resistance Heat

Electric resistance heating converts nearly 100 percent of the energy in the electricity to heat. However, most electricity is produced from oil, gas, or coal generators that convert only about 30 percent of the fuel's energy into electricity. Because of electricity's generation and transmission losses, electric heat is often more expensive than heat produced in the home with combustion appliances such as natural gas, propane, and oil furnaces.

Electric resistance heat can be supplied by centralized forced-air furnaces or by zonal heaters in each room, both of which can be composed of a variety of heater types. Zonal heaters distribute electric resistance heat more efficiently than electric furnaces because you set room temperatures according to occupancy. In addition, zonal heaters have no ducts that can lose heat before it reaches the room. However, electric furnaces can accommodate central cooling more easily than zonal electric heating because the air conditioner can share the furnace's ducts. Electric resistance heat can be provided by electric baseboard heaters, electric wall heaters, electric radiant heat, electric space heaters, electric furnaces, or electric thermal storage systems.

Electric Baseboard Heaters

Electric baseboard heaters are zonal heaters controlled by thermostats located within each room. Baseboard heaters contain electric heating elements encased in metal pipes. The pipes, surrounded by aluminum fins to aid heat transfer, run the length of the baseboard heater's housing or cabinet. As air within the heater is warmed, it rises into the room, and cooler air is drawn into the bottom of the heater. Some heat also radiates from the pipe, fins, and housing. Baseboard heaters are usually installed underneath windows. There, the heater's rising warm air counteracts falling cool air from the cold window glass. Baseboard heaters are seldom located on interior walls because standard heating practice is to supply heat at the home's perimeter where the greatest heat loss occurs. Baseboard heaters should sit at least three-quarters of an inch (1.9 centimeters) above the floor or carpet, to allow the cooler air on the floor to flow under and through the radiator fins so it can be heated. The heater should also fit tightly to the wall to prevent the warm air from circulating behind it and streaking the wall with dust particles.

The quality of baseboard heaters varies considerably. Cheaper models can be noisy and often give poor temperature control.

Electric Wall Heaters

Electric wall heaters consist of an electric element with a reflector behind it to reflect heat into the room and usually a fan to move air through the heater. They are usually installed on interior walls because installing them in an exterior wall makes that wall difficult to insulate.

Electric Radiant Heat

Electric furnaces and baseboard heaters circulate heat by moving air. In contrast, radiant heating systems radiate heat to the room's objects, including its people. For example, you can feel a ceiling-mounted radiant heating panel warming your head and shoulders if you stand underneath it.

There are several types of electric radiant heaters. The most common are electric heating cables imbedded in floors or ceilings. Other radiant heating systems use special gypsum ceiling panels equipped with factory-imbedded heating cables. Newer ceiling-mounted radiant panels made of metal provide radiant heat faster than other types because they contain less material to warm up. Radiant heat offers draft-free heating that is easily zoned. Unlike other heating systems, it occupies no interior space, thus allowing you complete freedom to place furniture without worrying about impeding air flow from floor registers or baseboard heaters.

Manufacturers claim that radiant heat can provide comfort similar to other systems at lower indoor air temperatures, saving around 5 percent of space heating costs. Critics of radiant heat say that it can be difficult to control air temperature with a thermostat. The large heat-storage capacity of the concrete or plaster surrounding the heating cables may result in greater-than-normal fluctuations in the room air temperature, since it takes quite a while to heat up the storage mass. Also, some occupants complain about their heads being too warm in rooms that utilize ceiling radiant heat.

Supplying heat at the ceiling or floor, which are locations that typically border the outdoors or unheated spaces, can result in greater heat losses. For example, if there are any flaws in a heated concrete slab or gaps in the ceiling insulation above heating elements, a significant percent of the electric heat may escape to the outdoors without ever heating the home.

Electric Space Heaters

Electric space heaters come in a wide variety of models, either built in or portable. These heaters may have fans to circulate heated air, and they may also be designed to transfer some of their heat by radiation. All of these heaters must be given adequate clearance to allow air to circulate safely. Portable space heaters (Figure 8.9), as well as many built-in space heaters for small rooms, have built-in thermostats. Larger rooms heated with built-in electric space heaters should have low-voltage thermostats installed in an area that maintains the room's average temperature.

Fireplaces

Fireplaces are very commonly used in family rooms and other living areas to give a warm and cozy feeling. These fireplaces can be wood or natural-gas fired. Generally, fireplaces transfer the heat by radiation, and hot combustion gases

Figure 8.9

An electric radiant space heater.

(carrying a lot of thermal energy) go out through the stack. Hot gases are lighter and rise up the chimney; a natural suction created by this flow draws the heated warm air from the room. Most of the time, the warm air heated in the room by the main heating fuel is also drawn into the fireplace and goes up the chimney, resulting in a net loss of energy. It is estimated that about 75 percent of the heated air is lost through the chimney. However, many people still use fireplaces inefficiently (Figure 8.10).

Advantages and disadvantages of direct heating systems are discussed in Table 8.4.

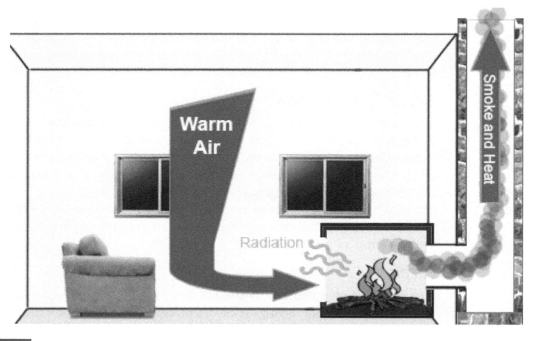

Figure 8.10

Heat loss through a fireplace.

Table 8.4
Advantages and Disadvantages of Direct Heating Systems

ADVANTAGES	DISADVANTAGES
▪ Generates heat at the point of use; no transmission losses ▪ Inexpensive to purchase and install ▪ Easy local control in each room ▪ In well-insulated houses it may be cheaper than other systems	▪ Heats only certain parts of the home ▪ Cannot be used for cooling ▪ Takes up living room ▪ Generally less efficient than other central heating systems

Cooling and Heating/Cooling Systems

Air conditioners and heat pumps are also forced-air systems. With these, cooled (and sometimes humidified or electronically cleaned) air is usually delivered through the same ductwork and registers used by heated air.

An air conditioner runs on electricity and removes heat from air with basic refrigeration principles. A heat pump can provide both heating and cooling. In the winter, a heat pump extracts heat from outside air and delivers it indoors. On hot summer days, it works in reverse, extracting heat from room air and pumping it outdoors to cool the house. Like air conditioners, nearly all heat pumps are powered by electricity. They have an outdoor compressor/condenser unit that is connected with refrigerant-filled tubing to an indoor air handler. As the refrigerant moves through the system, it completes a basic refrigeration cycle, warming or cooling the coils inside the air handler. The blower pulls in room air, circulates it across the coils, and pushes the air through ductwork back into the rooms. When extra heat is needed on particularly cold days, supplemental electric-resistance elements kick on inside the air handler to add warmth to the air that is passing through.

Heat Movers

So far we have discussed systems in which a fuel is burned and heat is produced and delivered into the home. In these systems we buy all the energy and (depending on the system) we reject or lose some heat to the surroundings, reducing the efficiency. One of the ways in which we can improve the heating efficiency is to make use of the heat that is available outside, even on a cold winter day. On such a day, with outside temperature at 30°F, the air still has more energy compared to air at 10°F or 5°F. Air at any temperature above absolute zero (0°K or –273°C) will have energy. The higher the air temperature, the higher its energy content. This energy can be transferred to the interior. A device that moves the heat from a low temperature environment to a high temperature environment

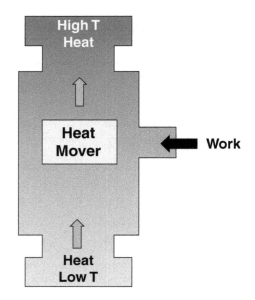

Figure 8.11

A schematic of heat flow in a heat mover.

is called a **heat mover** (Figure 8.11). This does not happen naturally or spontaneously, because heat always flows from high temperature to low temperature naturally. So some work needs to be done (energy spent) to move the heat from a low-temperature to a high-temperature environment. An example of this type of device is a heat pump. Do not get it confused with heat engines that we have seen in Chapter 3. A heat engine is a device that converts heat energy into mechanical energy. A heat pump does not convert energy. It just moves energy that is already there.

Heat Pumps

A **heat pump** is a device that moves heat from a low-temperature to a high-temperature environment with the help of work that is put in. The heat pumps are classified based on the low-temperature heat source:

> ❝A heat pump is a device that moves heat from a low-temperature to a high-temperature environment with the help of work that is put in.❞

1. *Air-source heat pump or air-to-air heat pump.* Heat is transferred from the low-temperature air outside to the high-temperature interior.

2. *Ground-source heat pump or ground-to-air heat pump.* The earth is used as a heat sink in the summer and a heat source in the winter; the pump relies on the relative warmth of the earth for its heating and cooling production.

3. *Water-source heat pump or water-to-air heat pump.* Heat is transferred from low-temperature water outside (from a pond or a lake) to a high-temperature interior.

A heat pump uses air-conditioning principles to extract heat from one place and deliver it to another, and vice versa. In addition to expelling heat from indoors, the system can be reversed to heat the home in the winter.

Air-to-Air or Air-Source Heat Pumps

The basic components and the operating principle of an air-source heat pump are illustrated in Figures 8.12 and 8.13. The main components of a heat pump are two sets of coils (tubes with large surface area), an expansion valve, and a compressor. A refrigerant is a substance that is easy to evaporate to form vapor when heat is absorbed, or to condense into liquid when heat is removed. The operation of a heat pump can be described as follows:

1. The refrigerant in the outside coils (in a liquid state) after the expansion valve absorbs heat from the air outside, because the refrigerant is much cooler than the air outside. During this process, the liquid refrigerant evaporates; hence the outside coils are referred to as "evaporator coils." The temperature of the refrigerant at this point is almost equal to the outside temperature.

2. This evaporated gas is compressed in the compressor to a higher temperature (so heat can be transferred to inside) and pressure. The compressor is

Heat Pump Heating Cycle

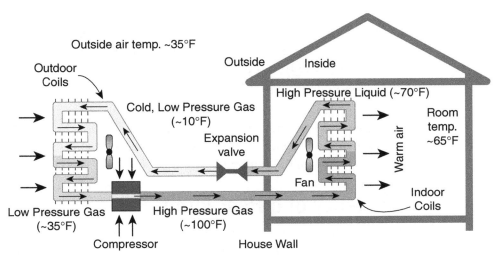

| Figure 8.12 | *Illustration of the operation of a heat pump in a heating cycle.* |

Heat Pump Cooling Cycle

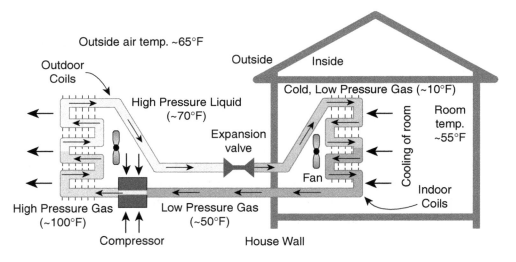

Outside air temp. ~65°F

Outdoor Coils

Outside   Inside

High Pressure Liquid (~70°F)

Cold, Low Pressure Gas (~10°F)

Expansion valve

Room temp. ~55°F

Cooling of room

Fan

Indoor Coils

High Pressure Gas (~100°F)

Low Pressure Gas (~50°F)

Compressor   House Wall

Figure 8.13

Illustration of the operation of a heat pump in a cooling cycle.

often thought of as the "heart" of the heat pump, since it does most of the work of forcing heat "uphill."

3. The high-temperature and high-pressure refrigerant passes through the indoor coil where the refrigerant gives up its heat to the indoor air. A fan blows air past the indoor coil to distribute heat to the house. This process cools the refrigerant to the point where much of it condenses, forming a liquid. Therefore, in the heating season, the indoor coil is called the "condenser coil." This change of state results in a large transfer of heat energy.

4. The warm refrigerant liquid now goes to the expansion valve. This device reduces the pressure, causing the refrigerant to become cold again—cold enough so that it is once again ready to absorb heat from the cool outdoor air and repeat the cycle.

Figure 8.13 shows the air source heat pump operating in a cooling cycle. The principle and the equipment are the same, except for the direction of flow. The heat is being picked up from inside at a cooler temperature and transferred outside at a much higher temperature.

Efficiency of a Heat Pump

The efficiency of a heat pump is measured using the term **Coefficient of Performance (COP)**. It is the ratio of useful heat that is pumped to a higher temperature to the unit amount of work that is put in. A general expression for the efficiency of a heat engine can be written as:

$$COP = \frac{Heat\ Energy_{hot}}{Work}$$

Using the same logic that was used for heat engines, this expression becomes:

$$COP = \frac{Q_{Hot}}{Q_{Hot} - Q_{Cold}} \qquad (8.1)$$

where Q_{Hot} = heat input at high temperature, and Q_{Cold} = heat rejected at low temperature. The expression can be rewritten as:

$$COP = \left(\frac{T_{Hot}}{T_{Hot} - T_{Cold}} \right) \qquad (8.2)$$

Illustration 8.1 shows that for every watt of power we use (and pay for) to drive this ideal heat pump, 13.3 W is delivered to the interior of the house and 12.3 from the outside (we don't pay for this). This seems to be a deal that one cannot refuse. However, the theoretical maximum is never achieved in the real world. In practice, a COP in the range of 2 to 6 is typical. Even with this range, it is an excellent choice, because for every watt of power that we use we transfer 1 to 5 additional watts from outside.

Illustration 8.1

Calculate the ideal coefficient of performance (COP) for an air-to-air heat pump used to maintain the temperature of a house at 70°F when the outside temperature is 30°F.

Solution: Using Equation 8.2, the COP can be calculated. However, the temperatures have to be converted to Kelvin. First the Fahrenheit scale should be converted to Celsius scale using:

$$°C = (°F - 32) \times \frac{5}{9}$$

T_{Hot} = 70°F = 21°C = 21 + 273 = 294 K
T_{Cold} = 30°F = −1°C = −1 + 273 = 272 K

$$COP = \frac{294\ K}{294\ K - 272\ K} = \frac{294}{22} = 13.3$$

Illustration 8.2

Compare the ideal coefficients of performance of the same heat pump installed in State College, PA and Ann Arbor, MI when the inside temperature of a house is maintained at 70°F at both locations and the outside temperatures on a given day were 40°F and 15°F at State College and Ann Arbor, respectively.

State College, PA	Ann Arbor, MI
$T_{Hot} = 70°F = 21°C = 294\ K$	$T_{Hot} = 70°F = 21°C = 294\ K$
$T_{Cold} = 40°F = 4°C = 277\ K$	$T_{Cold} = 15°F = -9.4°C = 264\ K$
$COP = \dfrac{T_{Hot}}{T_{Hot} - T_{Cold}} = \dfrac{294}{294 - 277}$	$COP = \dfrac{T_{Hot}}{T_{Hot} - T_{Cold}} = \dfrac{294}{294 - 264}$
$= 17.3$	$= 9.8$

During a heating season, the heat pump's efficiency increases on mild days and decreases on cold days.

Illustration 8.2 shows that when the outdoor temperature drops (in winter time) the efficiency or COP of an air-source heat pump decreases. The Heating Seasonal Performance Factor therefore is higher in a mild climate than in a region where winters are severe. While the Heating Seasonal Performance Factor of a heat pump will be lower in a cold winter area, the heat pump is generally more efficient than other electric heating systems in that area.

The Balance Point

As the outdoor temperature drops, the output of the heat pump decreases. At the same time, the heating requirement of the house increases. At some temperature the heat pump output and the home heating requirements match. This temperature is called the **balance point**, as shown in Figure 8.14. When the outdoor temperature falls below the balance-point temperature, supplemental heat will be required.

Ground Source (Geothermal) Heat Pumps

Ground-source (geothermal) heat pumps (GHPs) are similar to air-source heat pumps, except that the source of heat is the ground instead of outdoor air. The temperature of the earth below the frost line (that is generally 4–5 feet in Pennsylvania) remains constant throughout the year. The temperature is

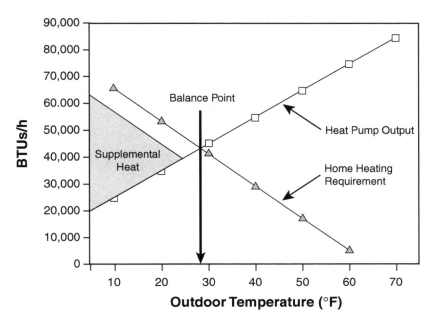

Figure 8.14

Balance point of a heat pump.

40–80°F, depending on the location. Ground source heat pumps use the earth as a heat sink in the summer and as a heat source in the winter, and therefore they rely on the relative warmth of the earth for their heating and cooling production. Through a system of underground pipes, they transfer heat from the warmer earth to the building in the winter, and they take the heat from the building in the summer and discharge it into the cooler ground. Therefore, GHPs do not *create* heat; they *move* it from one area to another.

Operating Principle

The GHP system operates much like an air-source heat pump, except that the outside coils are buried in the ground to extract or discharge the heat, with the addition of a few extra valves that allow heat-exchange fluid to follow two different paths: one for heating and one for cooling (Figure 8.15). The GHP takes heat from a warm area and exchanges the heat to a cooler area, and vice versa.

Classification of GHPs

The GHPs are classified as closed-loop and open-loop systems.

Closed-Loop Systems

Closed-loop systems are again of two types: horizontal and vertical loops. The horizontal type of installation is generally most cost-effective for residential

Geothermal Heat Pump

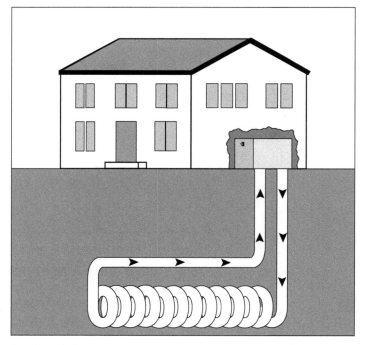

Geothermal heat pumps use the stable temperature of the shallow ground as a heat source to warm buildings in the winter and as a heat sink to cool them in the summer.

Figure 8.15

Ground-source heat pump.

installations, particularly for new construction where sufficient land is available (Figure 8.16). It requires trenches at least four feet deep.

The most common layouts either use two pipes, one buried at six feet and the other at four feet, or two pipes placed side by side at five feet in the ground in a two-foot wide trench. Or, the Slinky™ method of looping pipe allows more pipes in a shorter trench, which cuts down on installation costs and makes horizontal installation possible in areas not possible with conventional horizontal applications. Large commercial buildings and schools often use vertical systems because the land area required for horizontal loops would be prohibitive.

Vertical loops are also used where the soil is too shallow for trenching; they minimize the disturbance to existing landscaping (Figure 8.17). For a vertical system, holes (approximately four inches in diameter) are drilled about 20 feet apart and 100 to 400 feet deep. Into these holes go two pipes that are connected at the bottom with a U-bend to form a loop. The vertical loops are connected with horizontal pipe (i.e., manifold), placed in trenches, and connected to the heat pump in the building.

Figure 8.16

Horizontal-loop ground-source heat pump.
CREDIT: *Geothermal Heat Pump Consortium, Inc.*

Figure 8.17

A schematic of a vertical-loop ground-source heat pump.
CREDIT: *Geothermal Heat Pump Consortium, Inc.*

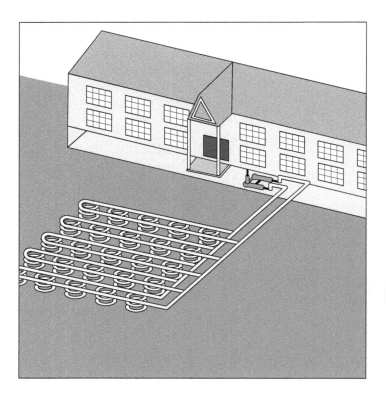

Figure 8.16

A schematic of a pond-loop system.
CREDIT: *Geothermal Heat Pump Consortium, Inc.*

Pond Closed Loops

If a home has source surface water, such as a pond or lake, this type of loop design may be the most economical, since there is no need to dig a trench or a well for the pipes in the ground. Figure 8.18 illustrates the arrangement of the loop. The fluid circulates through polyethylene piping in a closed system, just as it does in the ground loops. The pipe may be coiled in a slinky shape to fit more of it into a given amount of space. To assure sufficient heat-transfer capability, this loop is recommended only if the water level never drops below six to eight feet at its lowest level. Pond loops used in a closed system result in no adverse impacts on the aquatic system.

Open-Loop Systems

This type of system uses well(s) or surface body water as the heat exchange fluid that circulates directly through the GHP system. Once it has circulated through the system, the water returns to the ground through the well, a recharge well, or a surface discharge. This option is obviously practical only where there is an adequate supply of relatively clean water and when all local codes and regulations regarding groundwater discharge are met.

Factors Affecting the Type of GHP Loop

Geothermal heat pumps (GHPs) can be used effectively almost anywhere in the country. However, the specific geological, hydrological, and spatial characteristics of a site determine the best type of ground loop for a specific location.

1. **Geology.** Factors such as the composition and properties of your soil and rock (which can affect heat transfer rates) require consideration when designing a ground loop. For example, soil with good heat transfer properties requires less piping to gather a certain amount of heat than soil with poor heat transfer properties. The amount of soil available contributes to system design as well—system suppliers in areas with extensive hard rock or with soil too shallow to trench may install vertical ground loops instead of horizontal loops.

2. **Hydrology.** Ground or surface water availability also plays a part in deciding what type of ground loop to use. Depending on such factors as depth, volume, and water quality, bodies of surface water can be used as a source of water for an open-loop system or as a repository for coils of piping in a closed-loop system. Ground water can also be used as a source for open-loop systems, provided the water quality is suitable and all ground water discharge regulations are met.

3. **Land Availability.** The amount and layout of your land, your landscaping, and the location of underground utilities or sprinkler systems also contribute to your system design. Horizontal ground loops (generally the most economical) are typically used for newly constructed buildings with sufficient land. Vertical installations or more compact horizontal Slinky™ installations are often used for existing buildings because they minimize the disturbance to the landscape.

Benefits of a GHP System

Low Energy Use

The biggest benefit of GHPs is that they use 25 to 50 percent less electricity than conventional heating or cooling systems. This translates into a GHP using one unit of electricity to move three units of heat from the earth.

According to a report by Oak Ridge National Laboratory, statistically valid findings show that the 4,003-unit GHP retrofit project at Fort Polk, LA saves 25.8 million kilowatt-hours (kWh) in a typical year, or 32.5 percent of the pre-retrofit whole-community electrical consumption. This translates to an average annual savings of 6,445 kWh per housing unit. In addition, 100 percent of the whole-community natural gas previously used for space conditioning and water heating (260,000 therms) will be saved. In housing units that were entirely electric in the pre-retrofit period, the GHPs were found to save about 42 percent of the pre-retrofit electrical consumption for heating, cooling, and water heating.

Free or Reduced-Cost Hot Water

Unlike any other heating and cooling system, a geothermal heat pump can provide free hot water. A device called a "desuperheater" transfers excess heat from the heat pump's compressor to the hot water tank. In the summer, hot water is provided free; in the winter, water heating costs are cut roughly in half.

Year-Round Comfort

While producing lower heating bills, geothermal heat pumps are quieter than conventional systems and improve humidity control. These features help explain why customer surveys regularly show high levels of user satisfaction, usually well over 90 percent.

Design Features

Geothermal heat pump systems allow for design flexibility, and they can be installed in both new and retrofit situations. Because the hardware requires less space than that needed by conventional HVAC systems, the equipment rooms can be greatly scaled down in size, freeing space for productive use. Also, geothermal heat pump systems usually use the existing ductwork in the building and provide simultaneous heating and cooling without the need for a four-pipe system.

Improved Aesthetics

Architects and building owners like the design flexibility offered by GHPs. Historic buildings like the Oklahoma State Capital and some Williamsburg, Virginia structures use GHPs because they are easy to use in retrofit situations and easy to conceal, as they don't require cooling towers. GHP systems eliminate conventional rooftop equipment, allowing for more aesthetically pleasing architectural designs and roof lines. The lack of rooftop penetrations also means less potential for leaks and ongoing maintenance, as well as better roof warranties. In addition, the aboveground components of a GHP system are inside the building, sheltering the equipment both from weather-related damage and from potential vandalism.

Low Environmental Impact

Because a GHP system is so efficient, it uses a lot less energy to maintain comfortable indoor temperatures. This means that less energy—often created from burning fossil fuels—is needed to operate a GHP. According to the EPA, geothermal heat pumps can reduce energy consumption—and corresponding emissions—up to 44 percent compared to air-source heat pumps, and up to 72 percent compared to electric-resistance heating with standard air conditioning equipment.

Low Maintenance

According to a study completed for the Geothermal Heat Pump Consortium (GHPC), buildings with GHP systems had average total maintenance costs ranging from 6 to 11 cents per square foot, or about one third that of conventional systems. Because the workhorse part of the system—the piping—is underground or underwater, there is little maintenance required. Occasional cleaning of the heat exchanger coils and regularly changing the air filters are about all the work necessary to keep the system in good running order.

Zone Heating and Cooling

These systems provide excellent "zone" space conditioning. Different areas of the building can be heated or cooled to different temperatures simultaneously. For example, GHP systems can easily move heat from computer rooms (that need constant cooling) to the perimeter walls for winter heating in commercial buildings. School officials like the flexibility of heating or cooling just auditoriums or gymnasiums for special events rather than the entire school.

Durability

Because GHP systems have relatively few moving parts, and because those parts are sheltered inside a building, they are durable and highly reliable. The underground piping often carries warranties of 25 to 50 years, and the GHPs often last 20 years or more.

Reduced Vandalism

GHPs usually have no outdoor compressors or cooling towers, so the potential for vandalism is eliminated.

Installation and Operating Costs of GHP Systems

Typically when heating systems or appliances are compared, all the costs incurred—purchase, installation, operation, and maintenance costs—can be combined into a life-cycle cost, the cost of ownership over a period of years. Table 8.5 compares the various types of central heating systems.

On average, a geothermal heat pump (GHP) system costs about $2,500 to $3,500 per ton of capacity, or roughly $7,500 to 10,000 for a 3-ton unit (typical residential size). In comparison, other systems would cost about $4,000 with air conditioning (Figures 8.19 and 8.20). When included in the mortgage, the homeowner has a positive cash flow from the beginning. For example, say that the extra $3,500 will add $30 per month to each mortgage payment. A system using horizontal ground loops will generally cost less than a system with vertical loops. A 3,000 square-foot house in Oklahoma City has a verified average electric bill of $60 per month using a geothermal heat pump (Figure 8.19).

Table 8.5   *Comparison of Life Cycle Costs for Heat Pumps*

COMPARE	SAFETY	INSTALLATION COST	OPERATING COST	MAINTENANCE COST	LIFE-CYCLE COST
Combustion-based	A Concern	Moderate	Moderate	High	Moderate
Heat pump	Excellent	Moderate	Moderate	Moderate	Moderate
Geothermal or ground-source heat pump	Excellent	High	Low	Low	Low

Figure 8.19

A photo of a house using GHP in Oklahoma City.

SOURCE: *http://www.eere.energy.gov/geothermal/overview.html#heat_pump*

Energy costs about $1 per day for this 1,500-square-foot house that is heated and cooled with a GHP.

Figure 8.20

A picture of a home in Colorado that uses a GHP.

SOURCE: *http://www.eere.energy.gov/consumerinfo/heatcool/hc_space_geothermal_economics.html*

Geothermal heat pump installations in both new and existing homes can reduce energy consumption 25 to 75 percent compared to older or conventional replacement systems. Annual operating costs were also lowest with geothermal heat pumps. Add in the benefits of the desuperheater for hot water savings, and it's easy to see how a GHP system is the most efficient available.

Solar Energy for Home Heating

Energy that is received on the roof of a house is more than enough to supply the heating needs of the home. The energy reaching the Earth from the sun ranges from 600 to 2,000 BTUs per square foot per day (averaged over a year). It is a function of the latitude of the place. The amount of solar radiation reaching the earth is called the **insolation.** This is a short form for **in**cident **sol**ar radia**tion** per day.

The Earth revolves around the sun with its axis tilted toward the plane of rotation. In June, the North Pole is tilted toward the sun, and the solar rays are incident perpendicularly. Therefore, the sun appears to be at a higher angle (Figure 8.21). In December, the North Pole is tilted away from the sun, and therefore days are shorter and the solar rays are incident more obliquely, with lower energy flux (winter).

Figure 8.22 shows the position and angle of the sun during various times of the year.

Solar heating systems are classified as "active" or "passive" solar heating systems, or a combination of both.

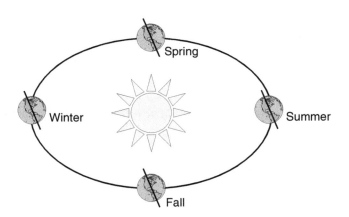

| Figure 8.21 | *Illustration of seasons for northern hemisphere.* |

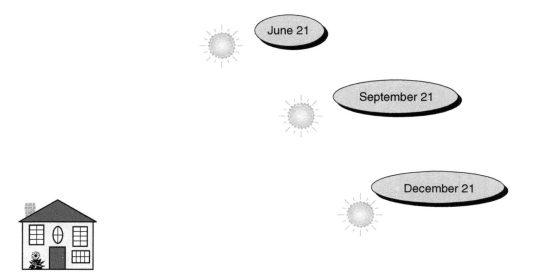

Figure 8.22

Position of the sun in the sky as a function of season.

Active Solar Heating Systems

Active solar heating systems consist of collectors that collect and absorb solar radiation, and electric fans or pumps to transfer and distribute the solar heat in a fluid (liquid or air) from the collectors. They generally have an energy storage system to provide heat when the sun is not shining. These methods basically have a collection device (collector), a storage device, and a distribution system. An active system often uses a flat plate collector mounted on the roof, which collects the solar energy and transfers it to a fluid (water or air). This transferred energy from the fluid is distributed directly or indirectly throughout the house. Figure 8.23 shows a schematic of an active solar heating system with a flat plate collector.

Flat plate collectors are usually placed on the roof or ground in the sunlight. The top or sunny side has a glass or plastic cover to let the solar energy in. The inside space is a black (absorbing) material to maximize the absorption of the solar energy. Cold water is drawn from the storage tank by pump #1 and is pumped through the flat plate collector mounted on the roof of the house. The water absorbs the solar energy and is returned to the tank. Warm water from the tank is pumped by pump #2 though the heating coil over which the air from the room is blown. The heated air then passes into the room and heats the room. Cold air sinks to the bottom and is recirculated over the heating coil. The standby electric coil is automatically turned on and provides the heat when the water temperature to the heating coil drops because of consecutive cloudy days.

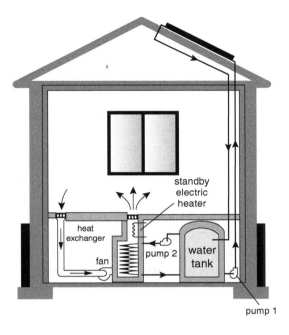

Figure 8.23

A schematic of an active solar heating system.

The efficiency of a collector can be expressed as:

$$Collector's\ Efficiency = \left(\frac{Useful\ energy\ delivered}{Insolation\ on\ collector} \right) \times 100\% \qquad (8.3)$$

Typical efficiencies range from 50–70 percent.

Passive Solar Heating Systems

Passive solar heating systems do not use mechanical devices such as fans, blowers, or pumps to distribute solar heat from a collector. Instead, they take advantage of natural heat flow to distribute warmth. An example of a passive system for space heating is a sunspace or solar greenhouse.

Passive systems also make use of materials with large heat capacities (stone, water, or concrete) to store and deliver heat. These are called thermal masses.

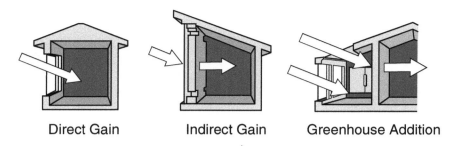

| Direct Gain | Indirect Gain | Greenhouse Addition |

Figure 8.24 *Illustration of direct gain, indirect gain, and solar greenhouse addition methods.*

The essential elements of a passive solar system are: solar collection (with south facing windows), thermal storage (concrete or stone walls), and excellent insulation (to prevent the loss of collected energy). Passive systems can be categorized into three types: direct gain, indirect gain, and attached sunspace and/or solar greenhouse. These systems are shown in Figure 8.24.

Direct gain systems allow the solar energy to come in through the south-facing window panes. Indirect gain systems allow the solar radiation to heat a wall; the energy is then slowly delivered into the interior of the house. The openings in the wall, as shown in Figure 8.24, promote convective currents. Cold room air enters the space between the glass panel and the wall through the bottom opening; as it gets heated, it rises to the top and comes in through the top opening. This is also called a Trombe wall. The figure also shows a greenhouse addition on the south side. This space is heated by the solar energy; a part of the energy is used for growing the plants, and some of it is used to heat the interior of the house.

Costs and Benefits

It is usually most economical to design an active system to provide 40 to 80 percent of the home's heating needs. Systems providing less than 40 percent of the heat needed for a home are rarely cost-effective except when using solar air heater collectors that heat one or two rooms and require no heat storage. A well-designed and insulated home that incorporates passive solar heating techniques will require a smaller and less costly heating system of any type, and it may need very little supplemental heat other than solar. The cost of an active solar heating system will vary. A simple window air heater collector can be made for a few hundred dollars. Installed commercial systems range from $30 and $80 per square foot of collector area. Usually, the larger the system, the less it costs per unit of collector area. Commercially available collectors come with warranties of 10 years or more and should easily last decades longer. Heating your home with an active solar energy system can significantly reduce your fuel bills in the winter. A solar heating system will also reduce the amount of air pollution and greenhouse gases.

Home Heating: Your "Power" in the Environmental Protection

Ways to Improve Energy Efficiency of Heating Systems

- Warm air, being less dense, rises through the chimney even if the fireplace is not on. Therefore, the dampers in the chimney must be tightly closed when the fireplace is not in use.

- Open fireplaces generally draw more warm air from the main room to support the burning of the fuel, and can become a drain on energy. Installation of glass doors reduces those convective losses. It also reduces the cold air infiltration.

- Combustion of fuel needs an adequate supply of air. This air normally comes from inside the house. Even with glass doors we need to provide vents for controlled amounts of air for combustion. It is more energy efficient if the air for combustion is drawn from outside and the inside vents are removed. The heat from the fireplace will be transferred into the room by radiation.

- Set your thermostat at the lowest possible comfortable temperature for you and your family. You can generate as much as a 12 percent savings by maintaining a 10-degree temperature setback.

- Use programmable thermostats, and program your clock thermostat to turn down the heat when no one is going to be home for four or more hours during the day.

- Replace your furnace filter at the beginning of winter and clean or replace it monthly during the heating season.

- If you have a hot water heating system, periodically "bleed" air trapped in your radiators to increase heat flow. Also, place a reflective material such as aluminum foil behind radiators located near walls, so heat will be reflected back into your home.

- If you have a forced-air heating system, seal the seams and joints in the ductwork with duct tape, and insulate them with vinyl-backed fiberglass insulation.

- If you have a gas-fired heating system, maximize its efficiency by getting it tuned up every year. Make sure the fan motor, burners, and circulator pump are properly maintained.

- If you have a heat pump, keep the outdoor unit clear of grass, leaves, and debris around the outdoor coil.

- Use a humidifier to make the air feel warmer so you can lower your heating thermostat.

- Close off rooms that aren't in use.

- Keep windows and doors near your thermostat tightly closed.

- Move furniture, draperies, and carpeting away from electric baseboards and registers.

- Seal seams and joints in ductwork with duct tape, and insulate them with vinyl-backed fiberglass insulation.

- Close drapes or cover windows at night and on cold, cloudy days. Keep draperies and shades open on sunny days.

- Proper insulation in walls, ceilings, and floors will significantly reduce heat loss in your home.

- Caulk and weatherstrip cracks in walls and floors, windows and walls to save energy and money.

- Keep your fireplace damper closed tightly when not in use.

- Close outside doors as quickly as possible. Just a few seconds with the door open lets in a lot of cold air.

Sources

Aubrecht, G. L. (1995). *Energy.* Englewood Cliffs, NJ: Prentice Hall.

Christensen, J. W. (1996). *Global science: Energy resources environment* (4th ed.). Dubuque, IA: Kendall/Hunt.

Fay, J. A., and Golomb, D. S. (2002). **Energy and the environment.** New York: Oxford University Press.

Hinrichs, R. A. (1992). *Energy.* Philadelphia: Saunders College Publishers.

http://www.eia.doe.gov/emeu/recs/recs2001/hc_pdf/spheat/hc3-1a_climate2001.pdf

http://www.hometips.com/cs-protected/guides/forcedair.html

http://www.bobvila.com/ArticleLibrary/Task/Building/ForcedAir.html

http://www.hometips.com/images/hyhw/comfort/38.gif

http://www.eere.energy.gov/consumerinfo/

http://www.oru.com/energyandsafety/energyefficiency/homeheatingsystems.html

questions

1. Compare and contrast the three conventional heating methods (forced air duct, hydronic floor, and baseboard heating systems).

2. Describe any five methods (total) to reduce home heating/cooling costs and explain how each step reduces the energy consumption.

3. Explain the operating principle of a ground source heat pump. Explain the difference between an open-loop and a closed-loop GHP.

4. What is the difference between active and passive heating systems?

5. With a neat sketch, describe how an active solar heating system works.

6. Describe any three passive heating methods.

7. Explain how a geothermal heat pump works and why it is so efficient.

8. Explain clearly how an air-to-air heat pump works. List the main components used in the heat pump.

multiple choice questions

1. What conversion takes place in a heat pump?
   a. Thermal energy to mechanical energy
   b. Chemical energy to mechanical energy
   c. Both conversions
   d. No conversion

2. A heat mover
   a. Uses work to move heat from a high-temperature reservoir to a low-temperature reservoir.
   b. Uses heat moving from a low-temperature reservoir to a high-temperature reservoir to produce work.
   c. Uses work to move heat from a low-temperature reservoir to a high-temperature reservoir.
   d. Uses heat moving from a high-temperature reservoir to a low-temperature reservoir to produce work.

3. A freezer is an example of a
   a. Heat generator
   b. Heat engine
   c. Heat pump
   d. All the above

4. Air-to-air heat pumps are _____ efficient than ground source heat pumps.
   a. More
   b. Less

5. Forced-air furnace systems distribute heat by
   a. Conduction
   b. Convection
   c. Radiation
   d. None of the above

6. Geothermal heat pumps
   a. Provide free or reduced cost hot water
   b. Require less maintenance
   c. Save energy and environment
   d. All the above

7. Water from an aquifer is used to extract energy and then discharge it back into the aquifer in a/an
   a. Horizontal-loop geothermal heat pump
   b. Vertical-loop heat pump
   c. Closed-loop heat pump
   d. Open-loop heat pump

8. If a heat pump is used to heat a house in the winter, how will the Coefficient of Performance be affected if the outside temperature drops?
   a. It increases
   b. It decreases
   c. It stays the same
   d. It may become negative

9. Flat plate collectors are usually placed on the roof or ground in the sunlight. They are classified as
   a. Active systems
   b. Passive systems
   c. Hybrid systems

10. Which of the following is not an advantage of forced air heating systems?
    a. Rooms are heated quickly
    b. The system isn't very complicated
    c. Ducts can easily be used for A/C in the summer
    d. They usually have a very high efficiency (99%)

11. Passive solar systems use
    a. Pumps
    b. Fans
    c. Motors
    d. None of the above

12. One of the heating systems is a radiant floor heating system. The space is mostly heated by
    a. Radiation
    b. Conduction
    c. Convection
    d. None of the above

13. Which of the following items is used in a passive solar heating system?
    a. Flat plate collector
    b. Photovoltaic cells
    c. Deciduous trees
    d. A geothermal heat pump

14. Hydronic radiant floor systems heat
    a. Localized spots
    b. Using air from outside
    c. The entire floor
    d. Some corners of the room

15. The most commonly used fuel used for residential heating is
    a. Coal   c. Gas
    b. Oil   d. Electric

16. When using a fireplace, most of the heated air is lost through the chimney.
    a. True
    b. False

17. A heat mover speeds up a natural heating process.
    a. True
    b. False

18. A "desuperheater" can provide a homeowner with
    a. air conditioning
    b. a filtrated heating system
    c. forced air heat
    d. free or reduced cost hot water

19. The compressor of a heat pump system _____ the pressure of the refrigerant, and _____ its temperature.
    a. decreases, lowers
    b. increases, raises
    c. increases, lowers
    d. decreases, raises

20. An example of a thermal mass is
    a. Wood   c. Stone
    b. Styrofoam   d. Straw

21. Gas and oil heating systems use the following rating for efficiency
    a. EER   c. AFUE
    b. COP   d. EF

22. Which term is used to define the energy efficiency of a central air-to-air heat pump?
    a. EER   c. EF
    b. HSPF   d. AFUE

23. Bleeding air from the pipes of a radiant baseboard heating system will
    a. Increase efficiency
    b. Increase amount of noise made by clanging pipes
    c. Decrease efficiency
    d. Improve air circulation in the house

problems for practice

1. A heat pump has a COP of 12. Explain the physical meaning of the number 12. The difference between T_{Hot} and T_{Cold} is 30°K. Calculate T_{Hot}.

2. A refrigerator operating in a house is maintained at 40°F, and the kitchen temperature happened to be 68°F. Calculate the COP of the refrigerator. If this refrigerator were moved to a garage, which is at 90°F, because the family purchased a new refrigerator for the kitchen, what would happen to its COP? Would it be a good idea to move the additional refrigerator to the garage?

Home Cooling

goals

- ☛ To understand the relationship between humidity and temperature

- ☛ To understand how an air conditioner works

- ☛ To gain familiarity with types of air conditioning systems

- ☛ To calculate the monetary savings when the efficiency of an air conditioner is improved

$\mathcal{A}$ir conditioning (A/C) involves cooling and heating air, cleaning it, and controlling its moisture level or humidity to provide maximum indoor comfort. An air conditioner transfers heat energy from the inside of a room or multiple rooms in a building to the outside. Outside air is NOT passing through the air conditioner, getting cooled and supplied inside. Outside air does not come in through the A/C. Only heat energy is moved or pumped from a low temperature environment (inside) to a high temperature environment (outside). Refrigerant in the system absorbs the excess heat and is pumped through a closed system of piping to an outside coil. A fan blows outside air over the hot coil, transferring heat from the refrigerant to the outdoor air. Because the heat is removed from the indoor air, the indoor area is cooled.

How Do We Measure Humidity?

Air contains varying amounts of moisture in a gaseous or vapor form. The actual amount of moisture contained in air is referred to as its absolute humidity. Precisely,

$$Absolute\ humidity = \frac{Mass\ of\ water\ vapor\ (lb)}{Mass\ of\ dry\ air\ (lb)} \qquad (9.1)$$

Air is a mixture of several gases, including nitrogen, oxygen, and water vapor. The total air pressure exerted by a volume of air in a given container on that container is the sum of the individual (partial) pressures of these gases. The vapor pressure is the partial pressure of the water vapor.

The warmer air is the more moisture it can hold. So its moisture-holding capacity changes with temperature. A chart showing the moisture content of air at various temperatures is called a psychometric chart (shown in Figure 9.1). Therefore, Relative Humidity is the ratio of the amount of moisture in the air to the maximum amount of moisture the air can hold at a given temperature.

$$Relative\ humidity_{At\ a\ given\ temperature}$$

$$= \left(\frac{Amount\ of\ water\ vapor\ (lb)}{Maximum\ amount\ of\ water\ vapor\ air\ can\ hold} \right) \times 100_{At\ that\ temperature} \qquad (9.2)$$

Air is said to be saturated (at 100 percent relative humidity) when it contains the maximum amount of moisture possible at that specific temperature. Air holding half the maximum amount of moisture at a given temperature has a relative humidity of 50 percent.

When relative humidity reaches 100 percent or saturated, moisture can condense. For example, when warm air, which can hold more water vapor, mixes with the cold air, the temperature of the resulting air will drop. Mixed air (being

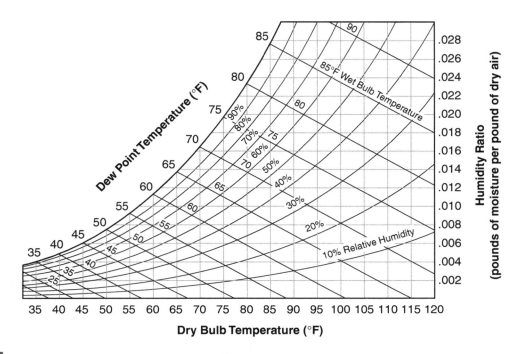

Figure 9.1

A simplified psychometric chart.
SOURCE: *http://hem.dis.anl.gov/eehem/95/951108.html*

a lower temperature than the warm air) cannot hold as much moisture, and, therefore, water vapor condenses or precipitation occurs. The temperature at which air reaches 100 percent relative humidity is called the dew point temperature. When the air is cooled below dew point, moisture in the air condenses.

Moisture will condense on a surface whose temperature is below the dew point temperature of the air next to it. For air at a given absolute humidity, the colder the surface, the higher the relative humidity next to that surface. So the coldest surface in a room is the place where condensation will probably occur first (called the first condensing surface).

Coming back to air conditioning, adjustment of humidity is important because we always try to cool warm air inside the room. When the temperature of the air decreases, the maximum amount of water the air can hold also decreases. So the relative humidity always increases. This is more conspicuous nn a hot humid day when the air conditioner is turned on and the air is cooled, relative humidity increases because the air cannot hold the same amount of water at lower temperatures Water can be seen dripping outside from an A/C. Relative Humidity is generally maintained at about 50 percent. Humidity that is too low or too high is very uncomfortable.

Illustration 9.1

Calculate the relative humidity of air when the air contains 0.002 lb of moisture per pound of dry air while the maximum moisture air can hold at that temperature is 0.005 lb per lb of dry air.

$$Relative\,Humidity = \frac{lb\,of\,moisture\,per\,lb\,of\,dry\,air}{Maximum\,lb\,of\,moisture\,per\,lb\,of\,dry\,air\,at\,that\,temp.}$$

$$= \frac{\dfrac{0.002\,lb}{lb\,of\,dry\,air}}{\left(\dfrac{0.005\,lb}{lb\,of\,dry\,air}\right)_{At\,that\,temp.}} \times 100 = 40\%$$

How an Air Conditioner Works

Figure 9.2 illustrates the principle of air conditioning. The compressor in your outdoor unit compresses the refrigerant (or "Freon") by transferring the part of the electrical energy it consumes into a high-temperature, high-pressure gas. As that gas flows through the outdoor coil, it loses heat to the surroundings (because the refrigerant is at a higher temperature than the outside temperature) and condenses into a high temperature, high pressure liquid. This liquid refrigerant travels through copper tubing into the evaporator coil. There the refrigerant expands. Its sudden expansion turns the refrigerant into a low temperature, low pressure gas. This cool gas then absorbs heat from the air in the room, which is blown over the evaporator coil. The cooled air is then distributed back through your room or multiple rooms. Meanwhile, the heat absorbed by the refrigerant is carried back outside through copper tubing and released into the outside air. So only the heat is transferred to the outside, not the air.

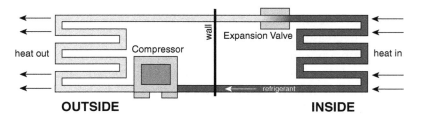

Figure 9.2   *Air conditioning principle.*

Types of Air Conditioners

The basic types of air conditioners are room air conditioners, split-system central air conditioners, and packaged central air conditioners.

Room Air Conditioners

Room air conditioners cool (Figure 9.3) rooms rather than the entire home. If they provide cooling only where they're needed, room air conditioners are less expensive to operate than central units, even though their efficiency is generally lower than that of central air conditioners.

Smaller room air conditioners (i.e., those drawing less than 7.5 amps of electricity) can be plugged into any 15- or 20-amp, 115-volt household circuit that is not shared with any other major appliances. Larger room air conditioners (i.e., those drawing more than 7.5 amps) need their own dedicated 115-volt circuit. The largest models require a dedicated 230-volt circuit.

Central Air Conditioners

Central air conditioners circulate cool air through a system of supply and return ducts. Supply ducts and registers (i.e., openings in the walls, floors, or ceilings covered by grills) carry cooled air from the air conditioner to the home. This

A room air conditioner.

SOURCE: : http://www.energystar.gov/ia/news/media_kit/images/roomac-sm.jpg

Figure 9.3

chapter 9   Home Cooling   **233**

cooled air becomes warmer as it circulates through the home; then it flows back to the central air conditioner through return ducts and registers. A central air conditioner is either a split-system unit or a packaged unit.

Split System

In a split-system central air conditioner as shown in Figure 9.4, an outdoor metal cabinet contains the condenser and compressor, and an indoor cabinet contains the evaporator. In many split-system air conditioners, this indoor cabinet also contains a furnace or the indoor part of a heat pump.

The air conditioner's evaporator coil is installed in the cabinet or main supply duct of this furnace or heat pump. If your home already has a furnace but no air conditioner, a split-system is the most economical central air conditioner to install. Most of the central air conditioners' systems are "split" systems.

Packaged Units

In a packaged central air conditioner, the evaporator, condenser, and compressor are all located in one cabinet, which usually is placed on a roof or on a concrete slab next to the house's foundation. This type of air conditioner also is used in small commercial buildings. Air supply and return ducts come from indoors

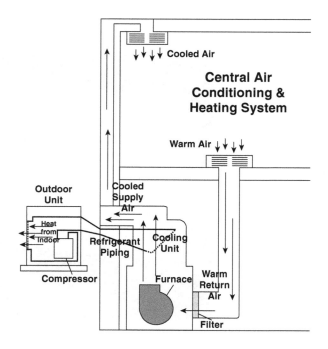

Figure 9.4

Illustration of a split air conditioner system.

through the home's exterior wall or roof to connect with the packaged air conditioner, which is usually located outdoors. Packaged air conditioners often include electric heating coils or a natural gas furnace. This combination of air conditioner and central heater eliminates the need for a separate furnace indoors.

Air Conditioner Efficiency

Each air conditioner has an energy-efficiency rating that lists how many BTUs per hour are removed for each watt of power it draws. For room air conditioners, this efficiency rating is the Energy Efficiency Ratio, or EER. For central air conditioners, it is the Seasonal Energy Efficiency Ratio, or SEER. These ratings are posted on an Energy Guide Label, which must be conspicuously attached to all new air conditioners. Energy Star-labeled appliances mean that they have high EER and SEER ratings.

In general, new air conditioners with higher EERs or SEERs have higher price tags. However, the higher initial cost of an energy-efficient model will be recovered several times during its lifespan. Some utility companies encourage the purchase of a more efficient air conditioner by offering incentives. Buy the most efficient air conditioner you can afford, especially if you use (or think you will use) an air conditioner frequently and/or if your electricity rates are high.

Room Air Conditioners—EER

Room air conditioners generally range from 5,500 BTUs per hour to 14,000 BTUs per hour. National appliance standards require room air conditioners built after January 1, 1990, to have an EER of 8.0 or greater. A room air conditioner with an EER of at least 9.0 is recommended for milder climates, whereas in hotter climates an EER over 10 is preferred.

The Association of Home Appliance Manufacturers reports that the average EER of room air conditioners rose 47 percent from 1972 to 1991. If a 1970s vintage room air conditioner with an EER of 5 is replaced with a new one with an EER of 10, air conditioning energy costs will be cut by 50 percent.

Central Air Conditioners—SEER

National minimum standards for central air conditioners require a SEER of 9.7 and 10.0 for single-package air conditioners and split systems, respectively. But you do not need to settle for the minimum standard—there is a wide selection of units with SEERs reaching nearly 17.

Before 1979, the SEERs of central air conditioners ranged from 4.5 to 8.0. Replacing a 1970s-era central air conditioner with a SEER of 6 with a new unit having a SEER of 12 will cut your air conditioning costs in half. Today's best air conditioners use 30 percent to 50 percent less energy to produce the same amount of cooling as air conditioners created in the mid 1970s. Even if your air conditioner is only 10 years old, you may save 20 to 40 percent of your cooling energy costs by replacing it with a newer, more efficient model.

Illustration 9.2 shows a calculation of power consumption of a room air conditioner.

Important Factors in Sizing Air Conditioners

Air conditioners are rated by the number of British Thermal Units (BTU) of heat they can remove per hour. Another common rating term for air conditioning size is the "ton," which is 12,000 BTUs per hour. The size of an air conditioner depends on:

- How large your home is and how many windows it has. Larger homes require more cooling, and more windows let more solar energy in.

- How much shade is on your home's windows, walls, and roof.

- How much insulation is in your home's ceiling and walls. Insulation resists heat gain.

- How much air leaks into your home from the outside.

- How much heat the occupants and appliances in your home generate.

Illustration 9.2

Calculate the power consumption of 5,000 BTUs/h room air conditioner with an Energy Efficiency Ratio (EER) of 8.

Solution:

$$EER = \frac{\frac{BTUs}{h} \, pulled \, out}{Watt}$$

We know that the AC pulls out 5,000 BTUs per hour and the EER = 8.

$$\text{Therefore, } \frac{5,000 \frac{BTUs}{h}}{Watts} = 8 \text{, or Power (Watts)} = \frac{5,000 \frac{BTUs}{h}}{8} = 625$$

Watts = 625

An air conditioner's efficiency, performance, durability, and initial cost depend on matching its size to the above factors. A system that is too large will cool the room or home quickly but not provide that comfort that is needed because the cool air reaches the thermostat quickly and the thermostat sends a signal to shut the system down before the relative humidity is reduced to a comfortable level. As the cold air is distributed in the room, the thermostat realizes that the temperature is not at the set point and then turns on the air conditioner. This quick cycling of the unit (start and stop) reduces the lifespan of the equipment and increases the energy consumption. A larger air conditioner also consumes more energy. A system that is small will have to work all the time and is not energy efficient. So the right size is very important for energy efficiency.

Determining the Size of Room Air Conditioners

1. Determine the square footage of the area to be cooled. For square and rectangular rooms, multiply the length of the area by its width.

2. Using the square footage and the chart below, determine the correct cooling capacity. Cooling capacity is measured in British Thermal Units (BTUs) per hour.

AREA TO BE COOLED (SQUARE FEET)	CAPACITY NEEDED (BTUs PER HOUR)
100 to 150	5,000
150 to 250	6,000
250 to 300	7,000
300 to 350	8,000
350 to 400	9,000
400 to 450	10,000
450 to 550	12,000
550 to 700	14,000
700 to 1,000	18,000
1,000 to 1,200	21,000
1,200 to 1,400	23,000
1,400 to 1,500	24,000
1,500 to 2,000	30,000
2,000 to 2,500	34,000

3. Make any adjustments for the following circumstances:
   - If the room is heavily shaded, reduce capacity by 10 percent.
   - If the room is very sunny, increase capacity by 10 percent.
   - If more than two people regularly occupy the room, add 600 BTUs for each additional person.
   - If the unit is used in a kitchen, increase capacity by 4,000 BTUs.
   - Consider where you install the unit. If you are mounting an air conditioner near the corner of a room, look for a unit that can send the airflow in the right direction.

Energy Savings by Naturally Cooling Your Home

Keeping cool indoors when it is hot outdoors is a problem. The sun beating down on the home causes indoor temperatures to rise to uncomfortable levels. Air conditioning provides some relief. But the initial costs of installing an air conditioner and the electricity costs to run it can be high. In addition, conventional air conditioners use refrigerants made of chlorine compounds, suspected contributors to the depletion of the ozone layer and to global warming. But there are alternatives to air conditioning.

Staying Cool

An alternative way to maintain a cool house or to reduce air-conditioning use is natural (or passive) cooling. Passive cooling uses non-mechanical methods to maintain a comfortable indoor temperature.

Specific methods to prevent heat gain include reflecting heat (i.e., sunlight) away from your house, blocking the heat, removing built-up heat, and reducing or eliminating heat-generating sources in your home.

Reflecting Heat Away

Dull, dark-colored home exteriors absorb 70 to 90 percent of the radiant energy from the sun that strikes the home's surfaces. Some of this absorbed energy is then transferred into a home by way of conduction, resulting in heat gain. In contrast, light-colored surfaces effectively reflect most of the heat away from a home.

Roofs

About one third of the unwanted heat that builds up in a home comes in through the roof. This is hard to control with traditional roofing materials. For example, unlike most light-colored surfaces, even white asphalt and fiberglass shingles absorb 70 percent of the solar radiation.

One good solution is to apply a reflective coating to your existing roof. Two standard roofing coatings are available. They have both waterproof and reflective properties and are marketed primarily for mobile homes and recreational vehicles.

Walls

Wall color is not as important as roof color, but it does affect heat gain somewhat. White exterior walls absorb less heat than dark walls. And light, bright walls increase the longevity of siding, particularly on the east, west, and south sides of the house.

Windows

Roughly 40 percent of the unwanted heat that builds up in your home comes in through windows. Reflective window coatings are one way to reflect heat away from your home. These coatings are plastic sheets treated with dyes or thin layers of metal. Besides keeping your house cooler, these reflective coatings cut glare and reduce the fading of furniture, draperies, and carpeting.

Two main types of coatings include sun-control films and combination films. Sun-control films are best for warmer climates because they can reflect as much as 80 percent of the incoming sunlight. Many of these films are tinted, however, and tend to reduce light transmission as much as they reduce heat, thereby darkening the room.

Combination films allow some light into a room, but they also let some heat in and prevent interior heat from escaping. These films are best for climates that have both hot and cold seasons. Investigate the different film options carefully to select the film that best meets your needs.

Blocking the Heat

Two excellent methods to block heat are insulation and shading. Insulation helps keep your home comfortable and saves money on mechanical cooling systems such as air conditioners and electric fans. Shading devices block the sun's rays and absorb or reflect the solar heat.

Insulation

Weatherization measures—such as insulating, weather stripping, and caulking—help seal and protect your house against the summer heat in addition to keeping out the winter cold.

The attic is a good place to start insulating because it is a major source of heat gain or loss. Adequately insulating the attic protects the upper floors of a house. Recommended attic insulation levels depend on where you live and the type of heating system you use. For most climates, you want a minimum of R-30. In climates with extremely cold winters, you may want as much as R-49.

Wall insulation is not as important for cooling as attic insulation because outdoor temperatures are not as hot as attic temperatures. Attic temperatures can reach as high as 150–160°F in summertime. Also, floor insulation has little or no effect on cooling.

Although unintentional infiltration of outside air is not a major contributor to inside temperature, it is still a good idea to keep it out. Outside air can infiltrate your home around poorly sealed doors, windows, electrical outlets, and through openings in foundations and exterior walls. Thorough caulking and weather stripping will control most of these air leaks.

Shading

Shading a home can reduce indoor temperatures by as much as 20°F. Effective shading can be provided by trees and other vegetation and exterior or interior shades.

Landscaping

Landscaping is a natural and beautiful way to shade your home and block the sun. A well-placed tree, bush, or vine can deliver effective shade and add to the aesthetic value of your property. When designing the landscaping, use plants that are native to the local area and that survive with minimal care.

Deciduous trees that lose their leaves in the fall help cut cooling energy costs the most. When selectively placed around a house, they provide excellent protection from the summer sun and permit winter sunlight to reach and warm your house. The height, growth rate, branch spread, and shape are all factors to consider in choosing a tree. Vines are a quick way to provide shading and cooling. Grown on trellises, vines can shade windows or the whole side of a house.

Besides providing shade, trees and vines create a cool microclimate that dramatically reduces the temperature (by as much as 9°F) in the surrounding area. During photosynthesis, large amounts of water vapor escape through the leaves, cooling the passing air. And the generally dark and coarse leaves absorb solar radiation.

Low ground cover such as grass, small plants, and bushes can also be very effective in cooling. A grass-covered lawn is usually 10°F cooler than bare ground in the summer. If you are in an arid or semiarid climate, consider native ground covers that require little water.

Shading Devices

Both exterior and interior shades control heat gain. Exterior shades are generally more effective than interior shades because they block sunlight before it enters windows. When deciding which devices to use and where to use them, consider whether you are willing to open and close them daily or just put them up for the hottest season. You also want to know how they will affect ventilation.

Exterior shading devices include awnings, louvers, shutters, rolling shutters and shades, and solar screens. Awnings are very effective because they block direct sunlight. They are usually made of fabric or metal and are attached above the window and extend down and out. A properly installed awning can reduce heat gain up to 65 percent on southern windows and 77 percent on eastern windows. A light-colored awning does double duty by also reflecting sunlight.

Shutters are movable wooden or metal coverings that, when closed, keep sunlight out. Shutters are either solid or slatted with fixed or adjustable slats. Besides reducing heat gain, they can provide privacy and security. Some shutters help insulate windows when it is cold outside.

Solar screens resemble standard window screens except they keep direct sunlight from entering the window, cut glare, and block light without blocking the view or eliminating air flow. They also provide privacy by restricting the view of the interior from outside your house. Solar screens come in a variety of colors and screening materials to complement any home. Although do-it-yourself kits are available, these screens will not last as long as professionally built screens.

Although interior shading is not as effective as exterior shading, it is worthwhile if none of the previously mentioned techniques are possible. There are several ways to block the sun's heat from inside your house.

Draperies and curtains made of tightly woven, light-colored, opaque fabrics reflect more of the sun's rays than they let through. The tighter the curtain is against the wall around the window, the more it will prevent heat gain. Two layers of draperies improve the effectiveness of the draperies' insulation when it is either hot or cold outside.

Venetian blinds, although not as effective as draperies, can be adjusted to let in some light and air while reflecting the sun's heat. Some newer blinds are coated with reflective finishes. To be effective, the reflective surfaces must face the outdoors.

Removing Built-Up Heat

Nothing feels better on a hot day than a cool breeze. Encouraging cool air to enter your house forces warm air out, keeping your house comfortably cool. However, this strategy only works when the inside temperature is higher than the outside temperature.

Natural ventilation maintains indoor temperatures close to outdoor temperatures and helps remove heat from your home. But only ventilate during the coolest parts of the day or night, and seal off your house from the hot sun and air during the hottest parts of the day.

The climate you live in determines the best ventilation strategy. In areas with cool nights and very hot days, let the night air in to cool your house. A well-insulated house will gain only 1°F per hour if the outside temperature is 85° to 90°F. By the time the interior heats up, the outside air should be cooler and can be allowed indoors.

In climates with daytime breezes, open windows on the side where the breeze is coming from and on the opposite side of the house. Keep interior doors open to encourage whole-house ventilation. If your location lacks consistent breezes, create them by opening windows at the lowest and highest points in your house. This natural "thermosiphoning," or "chimney," effect can be taken a step further by adding a clerestory or a vented skylight.

Illustration 9.3

What is the annual cost for operating a 3-ton central air conditioner with an SEER of 10? Assume that the AC operates 2,000 hours in a year and the cost of electricity is 9.2 cents per kWh.

Solution: Recall that 1 ton = 12,000 BTUs/h. Therefore, the cooling load is 3 × 12,000 BTUs/h = 36,000 BTUs/h.

$$SEER = \frac{\frac{BTUs}{h} \, pulled \, out}{Watt} = \frac{36,000 \frac{BTUs}{h}}{Watts} = 10$$

$$Watts = \frac{36,000 \frac{BTUs}{h}}{10} = 3,600W$$

Recall also that 1,000 W = 1 kW. Therefore, power consumption = 3.6 kW.

Energy = Power × Time of usage:

= 3.6 kW × 2,000 h/year = 7,200 kWh/year.

Annual cost = Units of energy × Price per unit:

$$= 7,200 \, kWh \times \frac{\$0.092}{kWh} = \$662.40$$

In hot, humid climates where temperature swings between day and night are small, ventilate when humidity is not excessive. Ventilating your attic greatly reduces the amount of accumulated heat, which eventually works its way into the main part of your house. Ventilated attics are about 30°F cooler than unventilated attics. Properly sized and placed louvers and roof vents help prevent moisture buildup and overheating in your attic.

Reducing Heat-Generating Sources

Often overlooked sources of interior heat gain are lights and household appliances, such as ovens, dishwashers, and dryers.

Because most of the energy that incandescent lamps use is given off as heat, use them only when necessary. Take advantage of daylight to illuminate your house. And consider switching to compact fluorescent lamps. These use about 75 percent less energy than incandescent lamps and emit 90 percent less heat for the same amount of light.

Illustration 9.4

If the owner bought an air conditioner with an SEER of 15, which costs $500 more, instead of the model in the previous Illustration 9.3, what is the payback period?

The power consumption of this new model is:

$$\frac{36,000 \frac{BTUs}{h}}{15} = 2,400 \text{ Watts} = 2.4 \text{ kW.}$$

Energy = 2.4 kWh × 2,000 h/year = 4,800 kWh/year.

$$\text{Annual cost} = \frac{4,800 \text{ kWh}}{\text{year}} \times \frac{\$0.092}{\text{kwh}} = \$441.60.$$

Savings per year = $662.40 − $441.60 = $220.80.

$$\text{Payback period} = \frac{\text{Additional investment}}{\text{Savings per year}} = \frac{\$500.00}{\$220.80/\text{year}} = 2.3 \text{ years.}$$

Many household appliances generate a lot of heat. When possible, use them in the morning or late evening when you can better tolerate the extra heat. Consider cooking on an outside barbecue grill or use a microwave oven, which does not generate as much heat and uses less energy than a gas or electric range.

Washers, dryers, dishwashers, and water heaters also generate large amounts of heat and humidity. To gain the most benefit, seal off your laundry room and water heater from the rest of the house.

Saving Energy

Using any or all of these strategies will help keep you cool. Even if you use air conditioning, many of these strategies, particularly reflecting heat and shading, will help reduce the energy costs of running an air conditioner. However, adopting all of these strategies may not be enough. Sometimes you need to supplement natural cooling with mechanical devices. Fans and evaporative coolers can supplement your cooling strategies and cost less to install and run than air conditioners. Ceiling fans make you feel cooler. Their effect is equivalent to lowering the air temperature by about 4°F. Evaporative coolers use about one-fourth the energy of conventional air conditioners but are effective only in dry climates.

Home Cooling: Your "Power" in the Environmental Protection

Central Air Conditioners

☛ Set your thermostat at 78°F or higher. Each degree setting below 78°F will increase your energy consumption by approximately 8 percent. The less difference there is between indoor and outdoor temperatures, the lower your overall cooling bill will be. And don't set your thermostat at a colder setting than normal when you turn on your air conditioner. It will not cool your home any faster and could result in excessive cooling and, therefore, unnecessary expense.

☛ Use bath and kitchen fans sparingly when the air conditioner is operating. These fans take out the cold air and make way for hot air to leak in.

☛ Inspect and clean both the indoor and outdoor coils. The indoor coil in your air conditioner acts as a magnet for dust because it is constantly wetted during the cooling season. Dirt buildup on the indoor coil is the single most common cause of poor efficiency. The outdoor coil must also be checked periodically for dirt buildup and cleaned if necessary.

☛ Check the refrigerant charge. The circulating fluid in your air conditioner is a special refrigerant gas that is put in when the system is installed. If the system is overcharged or undercharged with refrigerant, it will not work properly. You may need a service contractor to check the fluid and adjust it appropriately.

☛ Reduce the cooling load by using cost-effective conservation measures. For example, effectively shade east and west windows. When possible, delay heat-generating activities, such as dishwashing, until the evening on hot days. Turn off unused appliances

☛ Over most of the cooling season, keep the house closed tight during the day. Don't let in unwanted heat and humidity. If practical, ventilate at night either naturally or with fans.

☛ Try not to use a dehumidifier at the same time your air conditioner is operating. The dehumidifier will increase the cooling load and force the air conditioner to work harder.

Room Air Conditioners

☛ The unit should be leveled when installed, so that the inside drainage system and other mechanisms operate efficiently. If possible, install the unit in a shaded spot on your home's north or east side. Direct sunshine on the unit's outdoor heat exchanger decreases efficiency by as much as 10 percent. You can plant trees and shrubs to shade the air conditioner, but do not block the airflow.

☞ Don't place lamps or televisions near your air conditioner's thermostat. The thermostat senses heat from these appliances, which can cause the air conditioner to run longer than necessary.

☞ Set the fan speed on high except on very humid days. When humidity is high, set the fan speed on low for more comfort. The low speed on humid days will cool your home better and will remove more moisture from the air because of slower air movement through the cooling equipment. Consider using an interior fan in conjunction with your window air conditioner to spread the cooled air more effectively through your home without greatly increasing electricity use.

☞ Proper maintenance of your air conditioner will also save energy. Be sure to do the following:

■ At the start of each cooling season, inspect the seal between the air conditioner and the window frame to ensure it makes contact with the unit's metal case. Moisture can damage this seal, allowing cool air to escape from your house.

■ Check your unit's air filter once a month and clean or replace filters as necessary. Keeping the filter clean can lower your air conditioner's energy consumption by 5 percent to 15 percent.

■ Occasionally check the unit's drain channels. Clogged drain channels prevent a unit from reducing humidity, and the resulting excess moisture may discolor walls or carpet.

Sources

http://science.howstuffworks.com/ac.htm

Air Conditioning Contractors of America (ACCA). (1999). *Comfort, Air Quality, and Efficiency by Design,* Manual RS. 80 pp.

American Society of Heating, Refrigerating, and Air-Conditioning Engineers, Inc. (ASHRAE). (1998). *Cooling and Heating Load Calculation Principles.* 248 pp.

http://www.bobvila.com/ArticleLibrary/Subject/Special_Features/EnergyWise_House/ EEAirConditioning.html

U.S. Department of Energy report, CH10093-2221-FS 186. (1994). *Cooling Your Home Naturally.* 8 pp.

U.S. Department of Energy report, CH10099-379-FS 206. (1999). *Energy Efficient Air Conditioning.* 8 pp.

U.S. Department of Energy report, DOE/GO-10200101278-379-PS 228. (2001). *Cooling Your Home with Fans and Ventilation.* 8 pp.

http://www.energystar.gov/index.cfm?c=roomac.pr_properly_sized

questions

1. Define *relative humidity*.

2. Explain with a sketch the four main components of an air conditioner and their functions.

3. List the two main types of central air conditioner systems, and explain the difference between the two types.

4. What are the main factors to consider in sizing a room air conditioner?

5. What are natural (passive) ways in which you can reduce your cooling costs?

6. Define *EER* and *SEER*.

multiple choice questions

1. Which term is used to describe the efficiency of an air source heat pump when heating?
   a. AFUE
   b. EER
   c. EF
   d. HSPF

2. Oversized air conditioners (and heat pumps) do not run long enough to dehumidify the air.
   a. True
   b. False

3. Which term is used to define the energy efficiency of a room air conditioner?
   a. AFUE
   b. EER
   c. SEER
   d. COP

4. Each degree setting below 78°F will increase your energy consumption by approximately 8 percent.
   a. True
   b. False

5. In general, air conditioners are rated in tons. A ton of refrigeration transfers
   a. 6,000 BTUs/h
   b. 12,000 BTUs/h
   c. 24,000 BTUs/h
   d. 100,000 BTUs/h

6. Keeping the filter clean can lower your air conditioner's energy consumption by 5 to 15 percent.
   a. True
   b. False

7. _____ will change the amount of air conditioning required.
   a. The number of occupants
   b. The activity of the occupants
   c. The number of appliances operating the room
   d. All the above

8. Room air conditioners have a higher efficiency than central air conditioners.
   a. True
   b. False

9. Insulation of a house helps
   a. Heating costs
   b. Cooling costs
   c. Both heating and cooling costs

10. When warm outside air is cooled by an air conditioner, its relative humidity
    a. Increases
    b. Decreases
    c. Depends on the temperature
    d. Depends on the air conditioner

11. Air at 90°F contains 0.03 lb of moisture per lb of dry air. The saturation level at that temperature is 0.04/lb of dry air. What is the relative humidity?
    a. 25%
    b. 50%
    c. 75%
    d. 100%

12. The higher the "SEER" of a central air conditioner, the cheaper it is to operate.
    a. True
    b. False

13. The compressor is considered to be the "heart of the air conditioning system."
    a. True
    b. False

14. From an energy efficient point of view, an air conditioner should be mounted on
    a. The sunny side of the house
    b. The shady side of the house
    c. Any side if the A/C is efficient

15. A ton of air conditioning refers to the weight of the system.
    a. True
    b. False

16. Heat pumps can cool your home.
    a. True
    b. False

17. Keeping the filter clean can lower your air conditioner's energy consumption.
    a. True
    b. False

18. Roof overhangs reduce cooling costs.
    a. True
    b. False

19. When air is heated, the maximum amount of water vapor it can hold will
    a. Increase
    b. Decrease
    c. Remain constant

20. Air is saturated when
    a. It contains the maximum amount of moisture possible at that temperature
    b. It contains the minimum amount of moisture possible at that temperature
    c. Its relative humidity is 0%

21. When air is cooled below the dew point, moisture in the air
    a. Condenses
    b. Evaporates

22. _____ relative humidity feels uncomfortable.
    a. Very low
    b. Both very high and very low
    c. Very high

23. Cooling costs are _____ for the basement (in ground) compared to the upper floors.
    a. Higher
    b. Lower
    c. Same

problems for practice

1. A contractor says a home requires a 5-ton air conditioning system. How many BTUs does this system pull out each hour?

2. John Jankomsky has an air conditioner that is rated at 10,000 BTUs/h with an EER of 10. What is the power consumption of the air conditioner?

3. Your old high school pal Mike Errington wants to upgrade an old 1976 vintage room air conditioner that is believed to operate at an EER of 5.5. He is considering a room air conditioner with an EER of 11. He wants to know by what percentage electricity consumption would be reduced. Can you help him with the data given or not?

4. Suppose you are comparing two air conditioners both of which last for 10 years. The least efficient air conditioner draws 775 watts of power. The most efficient one uses 700 watts. Assuming that the air conditioner operates 2,000 hours annually and that the local energy costs 0.09 per kWh, how much money and energy can you save with the energy efficient model? If the energy efficient model costs $50 more than the least energy efficient model, would you buy the more energy efficient model? Justify your answer quantitatively.

Windows

goals

☞ **To explain how windows work**

☞ **To understand the mechanisms of heat loss through windows**

☞ **To know the important factors in selecting windows**

☞ **To be able to calculate payback period when replacing with energy-efficient windows**

indows typically occupy about 15 to 20 percent of the surface area of the walls. Windows not only add aesthetic looks and are often a very important aspect of a home, they also are a very significant component of home heating and cooling costs. Windows lose more heat per square foot of area in winter and gain more heat in summer than any other surface in the home. We already discussed in Chapter 7 that simple glass (⅛th inch) has a very low R-value (0.03). So even if the walls are well insulated to an R-value of about 13 to 19 but the windows have poor R-value, most of the heat escapes through the windows and the purpose of having a well insulated wall is lost. It is estimated that in 1990 alone, the energy used to offset unwanted heat losses and gains through windows in residential and commercial buildings cost the United States $20 billion (one-fourth of all the energy used for space heating and cooling). However, when properly selected and installed, windows can help minimize a home's heating, cooling, and lighting costs.

Figure 10.1 illustrates heat loss through a window. Heat always flows from high temperature to low temperature. When heated air in the room comes in contact with the cooler window pane, heat is lost through the window, and it gets cold. Cold air sinks to the bottom, and that stream of cold air hits a person and makes him uncomfortable. Although energy is spent to heat the air in the room, it will not be comfortable. Making the windows efficient can significantly save energy and money. Figure 10.2 shows the savings that can be obtained in Boston, a relatively heating intensive place.

Similarly, poor windows allow the solar energy to penetrate through the windows and heat the space. Incoming solar radiation, which consists of infrared

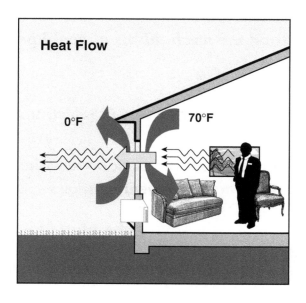

Figure 10.1

Heat and comfort loss through a window.

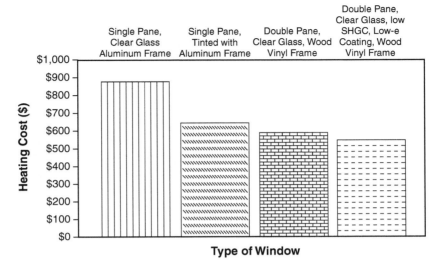

Figure 10.2   Savings in annual heating costs with energy efficient windows for a typical house in Boston, MA.

(IR), ultraviolet (UV), and visible waves, enters the room. The IR radiation, which is also called heat radiation, heats the space excessively and adds to the air conditioning in summertime as illustrated in Figure 10.3. Therefore, energy efficient windows are critical in summertime or even in places where the cooling requirement is high. Figure 10.4 shows the annual savings of an energy efficient window in Phoenix, Arizona during a cooling season.

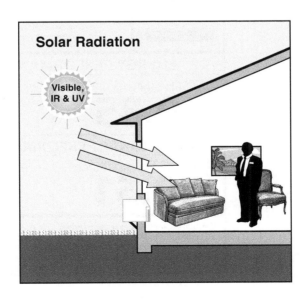

Figure 10.3   Solar heat gain in summer time through windows.

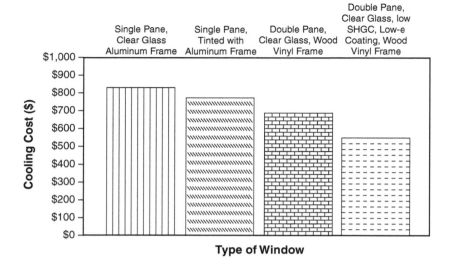

Figure 10.4   *Savings in annual cooling costs with energy efficient windows for a typical house in Phoenix, AZ.*

Figure 10.5   *National Fenestration Rating Council's window label.*

Source: *National Fenestration Rating Council*

The National Fenestration Rating Council (NFRC) developed an energy performance label (shown in Figure 10.5) that helps determine a window's performance. It indicates how much it helps to cool a building in the summer, to warm it in the winter, to keep out wind, and to resist condensation. By using the information contained on the label, builders and consumers can reliably compare one product with another and make informed decisions about the windows, doors, and skylights they buy. The NFRC adopted a new energy performance label in 1998. It lists the manufacturer, describes the product, provides a source for additional information, and includes ratings for one or more energy performance characteristics.

Factors in Window Selection

The NFRC rates all products in two standard sizes so that consumers and others can be sure they are comparing products of the same size. There are three factors that will be listed on the label with some additional information. These factors are the U-factor, Solar Heat Gain Coefficient (SHGC), and Visible Transmittance (VT).

U-Factor or Value

The U-factor measures how well a product prevents heat from escaping. The rate of heat loss is indicated in terms of the U-factor (U-value) of a window assembly. U-factor ratings generally fall between 0.20 and 1.20. The smaller the U-factor the less heat the window allows to pass. Most window manufacturers label their windows with a U-value (conductance of heat, BTU/h °F ft^2). U-values are the reciprocals of R-values (h °F ft^2/BTU). The lower the U-value, the less heat is lost through the window. The insulating value is indicated by the R-value, which is the inverse of the U-value. The lower the U-value, the greater a window's resistance to heat flow and the better its insulating value.

Some manufacturers rate thermal performance using R-value. R-value is the inverse of the U-factor, i.e., $1/U = R$, $1/R = U$. For example, a U-factor of 0.25 is the same as an R-factor of 4.0. The overall or "total" or "whole window" U-factor of any window depends on the type of glazing, the frame materials and size, glazing coatings, and type of gas (air, or inert argon or krypton) between the panes. Some typical U- and R-factor ranges for different window assemblies are shown in Table 10.1.

Solar Heat Gain Coefficient

The Solar Heat Gain Coefficient (SHGC) measures how well a window blocks heat from the sunlight. The SHGC is the fraction of incident solar radiation

Table 10.1 *Typical U-factors and R-values of Windows*

WINDOW ASSEMBLY	U-FACTOR	R-VALUE
Single Glazed	0.91–1.11	1.1–0.9
Double Glazed	0.43–0.57	2.3–1.7
Triple Glazed	0.15–0.33	6.7–3.3

admitted through a window, both directly transmitted, and absorbed and subsequently released inward. The SHGC is expressed as a number between 0 and 1. The lower a window's solar heat gain coefficient, the less solar heat it transmits.

Visible Transmittance

Visible Transmittance (VT) measures how much light comes through a window. The visible transmittance is an optical property that indicates the amount of visible light transmitted. VT is expressed as a number between 0 and 1. The higher the VT, the more light is transmitted.

The ratio between SHGC and VT is called the light-to-solar gain ratio (LSG.) This provides a gauge of the relative efficiency of different glass types in transmitting daylight while blocking heat gains. The higher the ratio number the brighter the room is without adding excessive amounts of heat.

Table 10.2 lists typical SHG, VT, and LSG values for the Total Window and Center of Glass (in parenthesis) for different types of windows.

The following two parameters are not required to be reported on NFRC labels but are optional.

Table 10.2 *Solar Heat Gain Coefficients, Visible Transmittance, and LSG Values for Various Types of Windows*

WINDOW AND GLAZING TYPES	SHG	VT	LSG
Single-glazed, clear	0.79 (0.86)	0.69 (0.90)	0.87 (1.04)
Double-glazed, clear	0.58 (0.76)	0.57 (0.81)	0.98 (1.07)
Double-glazed, bronze	0.48 (0.62)	0.43 (0.61)	0.89 (0.98)
Double-glazed, spectrally selective	0.31 (0.41)	0.51 (0.72)	1.65 (1.75)
Double-glazed, spectrally selective	0.26 (0.32)	0.31 (0.44)	1.19 (1.38)
Triple-glazed, new low-e	0.37 (0.49)	0.48 (0.68)	1.29 (1.39)

Air Leakage

Air Leakage (AL) is indicated by an air leakage rating expressed as the equivalent cubic feet of air passing through a square foot of window area (cfm/sq. ft). Heat loss and gain occur by infiltration through cracks in the window assembly by convection. The lower the AL, the less air will pass through cracks in the window assembly.

Condensation Resistance

Condensation Resistance (CR) measures the ability of a product to resist the formation of condensation on the interior surface of that product. The higher the CR rating, the better that product is at resisting condensation formation. While this rating cannot predict condensation, it can provide a credible method of comparing the potential of various products for condensation formation. CR is expressed as a number between 0 and 100.

Advances in Window Technologies

Before innovations developed in glass, films, and coatings in the past decade, a typical residential window with one or two layers of glazing allowed roughly 75 to 85 percent of the solar energy to enter a building. This has a negative impact on summertime comfort and cooling bills, especially in hot climates. The properties of a normal window are illustrated in Figure 10.6.

External window shading devices such as awnings, roof overhangs, shutters, and solar screens, along with internal shading devices such as curtains and blinds, can control the entry of solar heat. However, shutters, solar screens, curtains, and blinds make rooms dark. Curtains and blinds also let in some undesirable heat. While exterior shading devices are about 50 percent more effective than internal devices at blocking solar heat, they may create problems with the building's aesthetics and are sometimes expensive to build. It is also impractical to construct roof overhangs to effectively shade windows facing east and west. Table 10.3 shows the percentages of radiant energy that different types of internal shading devices transmit, reflect, or absorb.

Research in the recent past led to the development of low-emissivity or "low-e" glass and films that control heat gain and loss, reduce glare, minimize fabric fading, provide privacy, and occasionally provide added security in wind, seismic, and other high-hazard zones. New construction and window replacement applications commonly use glazing with these coatings.

Some low-e coatings and solar control films reduce solar heat gain without excessively impairing visible light transmission. These include tinted glass and spectrally selective coatings, which transmit visible light while reflecting the

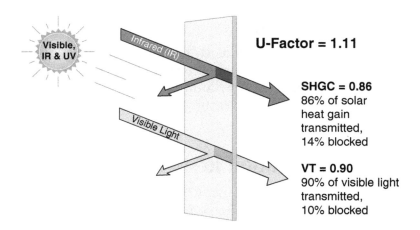

Figure 10.6

Typical properties of a normal window.

Table 10.3 *Percent Radiant Energy Transmitted, Reflected and Absorbed for Different Types of Shading Materials*

SHADE TYPE	TRANSMITTED ENERGY (PERCENT)	REFLECTED ENERGY (PERCENT)	ABSORBED ENERGY (PERCENT)
Roller Shades	25 percent	15–80 percent	20–65 percent
Vertical Blinds	0 percent	23 percent	77 percent
Venetian Blinds	5 percent	40–60 percent	35–55 percent

infrared portion of sunlight. Many spectrally selective coatings also have some low-e properties as well. Modern window glazing falls into three categories: chemically or physically altered glass, coated glass or films, and multiple-layered assemblies with or without either of the first two items.

Chemically or Physically Altered Glass

Tinting is the oldest of all the modern window technologies and, under favorable conditions, it can reduce solar heat gain during the cooling season by 25 to 55 percent. Tinted glass is made by altering the chemical properties of the glass. Both glass and plastic laminate may be tinted. The tints absorb a portion of the sunlight and solar heat before they can pass all the way through the window to the room. Tinted glazings reduce the latter by 25 to 55 percent. "Heat absorbing" tinted glass maximizes its absorption across some, or all, of the solar spectrum. Unfortunately, the absorbed energy often transfers by radiation and convection to the inside.

Spectrally Selective Coatings

Spectrally selective coatings or tints reduce infrared light (heat) transmission while allowing relatively more visible light to pass through (compared to bronze- or gray-tinted glass). For buildings that use daylight for lighting, a spectrally selective window is a good choice. Spectrally selective glass also absorbs much of the ultraviolet (UV) portion of the solar spectrum. In a multi-paned window, it functions best as the outermost sheet of glazing. Thermal performance is increased when it is combined with a low-e coating. Spectrally selective coatings often have a light blue or green tint.

Low-e Coatings

Low-e and reflective coatings usually consist of a layer of metal a few molecules thick. The thickness and reflectivity of the metal layer (low-e coating) and the location of the glass it is attached to directly affect the amount of solar heat gain in the room. Most window manufacturers now use one or more layers of low-e coatings in their product lines. Any low-e coating is roughly equivalent to adding an additional pane of glass to a window. Low-e coatings reduce IR heat transfer by 5 to 10 times. The lower the emissivity value (a measure of the amount of heat transmission through the glazing or coating), the more the material reduces the heat transfer from the inside to the outside. Most low-e coatings also slightly reduce the amount of visible light transmitted through the glazing relative to clear glass. Table 10.4 gives the emissivity values for different types of glass. There are three types of coating available: soft, hard coatings, and Heat Mirror.

Soft coat is applied to the surface of glass at lower temperatures. It's not durable enough to be exposed to the elements, so it's only used on the inner surfaces of windows that are not exposed to the elements. Hard coat is produced by fusing metallic oxide to the hot surface of glass during manufacture and is found primarily on storm windows and removable energy panels. Hard coat is applied on the glass surface at a high temperature. One layer is about 1/10,000th the diameter of a human hair. Hard coat is not quite as energy efficient as soft coat, but it is tough enough to be used on surfaces exposed to the elements. Both types of low-e coatings (within insulated glazing assemblies) typically last for 10 to 50 years.

Table 10.4   *Emissivity Values for Various Low-e Coatings*

TYPE OF COATING	EMISSIVITY
Clear glass; uncoated	0.84
Glass with single hard coat low-e	0.15
Glass with single soft coat low-e	0.10

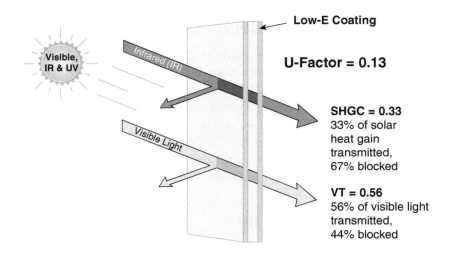

Low-E Coating

U-Factor = 0.13

Visible, IR & UV

Infrared (IR)

Visible Light

SHGC = 0.33
33% of solar
heat gain
transmitted,
67% blocked

VT = 0.56
56% of visible light
transmitted,
44% blocked

Figure 10.7

Typical properties of a low-e coated window.

Heat Mirror is a proprietary product that's applied to a thin polyester sheet suspended between the two panes of dual pane window. The coating reflects radiant heat while the sheet decreases heat loss by splitting the air space in two.

The only spectrally selective coatings now available are modified soft coat low-e coatings. The selective properties of the coatings are determined by modifying the coating's thickness and number of layers. A spectrally selective tinted glazing with a pyrolytic hard coat serves a similar purpose. These spectrally selective hard coats are currently under development.

"Aftermarket" films are available for application on existing windows. They are relatively easy to apply on glazing up 36 square inches. They are often applied to the glass with a water-soluble adhesive. To reduce the possibility of bubbles and wrinkles on large windows, have the film installed professionally. Most films should be applied to the inside surface of the glass since they can be damaged easily by weather. If you plan to install the film yourself, be careful to select the appropriate film for your needs and understand all directions before beginning. Plastic films generally last about 8 to 10 years before they start looking worn.

Gas Fills

Filling the space with a less conductive, more viscous, or slow-moving gas minimizes the convection currents within the space. In addition, conduction through the gas is reduced, and the overall transfer of heat between the inside and outside is reduced.

Argon and krypton gas with measurable improvement in thermal performance have been used.

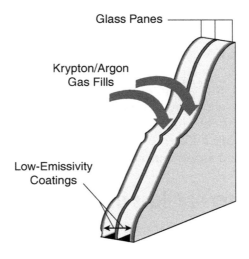

Glass Panes

Krypton/Argon
Gas Fills

Low-Emissivity
Coatings

Figure 10.8

Illustration showing glazing and gas filling.
SOURCE: *http://eetd.lbl.gov/success/window.html*

Gas-Filled Windows

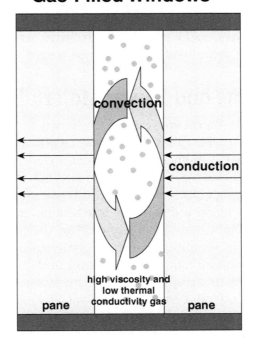

convection

conduction

high viscosity and
low thermal
conductivity gas

pane   pane

Figure 10.9

Illustration of a gas-filled window.

Argon gas filling provides an effective thermal resistance level of R-7 per inch, krypton gas provides R-12.5 per inch, and xenon gas provides R-20 per inch. Argon is inexpensive, nontoxic, nonreactive, clear, and odorless. The optimal spacing for an argon-filled unit is the same as for air, about ½ inch (11–13 mm). Krypton has better thermal performance, but it is more expensive to produce. Krypton is particularly useful when the space between glazings must be thinner than normally desired, for example, ¼ inch (6 mm). The optimum gap width for krypton is ⅜ inch (9 mm). A mixture of krypton and argon gases is also used as a compromise between thermal performance and cost.

Layers of Glass and Air Spaces

Standard single-pane glass has very little insulating value (approximately R-1). It provides only a thin barrier to the outside and can account for considerable heat loss and gain. Traditionally, the approach to improve a window's energy efficiency has been to increase the number of glass panes in the unit, because multiple layers of glass increase the window's ability to resist heat flow.

Double- or triple-pane windows have insulating air- or gas-filled spaces between each pane. Each layer of glass and air space resists heat flow. The width of the air spaces between the panes is important, because air spaces that are too wide (more than ⅝ inch) or too narrow (less than ½ inch) have lower R-values (i.e., they allow too much heat transfer). Advanced, multi-pane windows are now manufactured with inert gases (argon or krypton) in the spaces between the panes because these gases transfer less heat than air. Multi-pane windows are considerably more expensive than single-pane windows and limit framing options because of their increased weight.

Frame and Spacer Materials

Window frames are available in a variety of materials, including aluminum, wood, vinyl, and fiberglass. Frames may be primarily composed of one material, or they may be a combination of different materials such as wood clad with vinyl or aluminum-clad wood. Each frame material has its advantages and disadvantages.

Though ideal for strength and customized window design, aluminum frames conduct heat and therefore lose heat faster and are prone to moisture condensation. Through anodizing or coating, the corrosion and electro-galvanic deterioration of aluminum frames can be avoided. Additionally, the thermal resistance of aluminum frames can be significantly improved by placing continuous insulating plastic strips between the interior and exterior of the frame.

Wood frames have higher R-values, are not affected by temperature extremes, and do not generally promote moisture condensation. Wood frames, however, require considerable maintenance in the form of periodic painting or staining. If not properly protected, wood frames can swell, which leads to rot, warping, and sticking.

Vinyl window frames, which are made primarily from polyvinyl chloride (PVC), offer many advantages. Available in a wide range of styles and shapes, vinyl frames have moderate to high R-values, are easily customized, are competitively priced, and require very low maintenance. While vinyl frames do not possess the inherent strength of metal or wood, larger-sized windows are often strengthened with aluminum or steel reinforcing bars.

Fiberglass frames are relatively new and are not yet widely available. With some of the highest R-values, fiberglass frames are excellent for insulating and will not warp, shrink, swell, rot, or corrode. Unprotected fiberglass does not hold up to the weather and therefore is always painted. Some fiberglass frames are hollow while others are filled with fiberglass insulation.

Spacers

Spacers are used to separate multiple panes of glass within the windows. Although metal (usually aluminum) spacers are commonly installed to separate glass in multi-pane windows, they conduct heat. During cold weather, the thermal resistance around the edge of a window is lower than in the center; thus, heat can escape, and condensation can occur along the edges.

To alleviate these problems, one manufacturer has developed a multi-pane window using a ⅛-inch-wide (0.32 centimeters-wide) PVC foam separator placed along the edges of the frame. Like other multi-pane windows, these use metal spacers for support, but because the foam separator is secured on top of the spacer between the panes, heat loss and condensation are reduced. Several window manufacturers now sandwich foam separators, nylon spacers, and insulation materials such as polystyrene and rockwool between the glasses inside their windows.

In order to overcome the thermal inefficiency of conventional aluminum spacers, a new type of spacer product called warm-edge technology has evolved in the industry. Warm-edge refers to the type of spacer material used to separate the panes of glass (or glazing) in an insulated window unit. If the material conducts less heat or cold than a conventional aluminum spacer at the edge of the glass, it is said to be "warm-edge." Most of these newer spacers are less conductive and outperform pure aluminum. But, still, they all contain some kind of metal. And metal is highly conductive. A no-metal Super Spacer is available in the market that uses no metal and consists of 100 percent polymer structural foam. Therefore it is believed to improve the R-Value of the whole window and alleviate moisture condensation problems.

General guidelines for selecting the main parameters for windows based on the climate are provided in Table 10.5.

Figure 10.10 shows the improvement in window performance (R-value) with advanced window glazing. It can be seen from the figure that the super windows are losing less heat and are even making progress in gaining heat instead of losing heat because of solar heat gains

Table 10.5   *Recommended Minimum Values for Window Parameters*

PLACE	COLDER CLIMATE	MODERATE CLIMATE	WARM CLIMATE
U-Factor	Less than 0.33	0.33	0.33
Visual Transmittance	50 percent	>50 percent	>60 percent
SHGC	0.4–0.55	>0.55	<0.4
UV-Protection	>75 percent	75 percent	75 percent
Edge Spacers	Super spacers	Warm-edge spacers	Warm-edge spacers
Frame	Non-conductive	Non-conductive	Non-conductive
Air leakage	<0.3 cfm/sq. ft	<0.3 cfm/sq. ft	<0.3 cfm/sq. ft

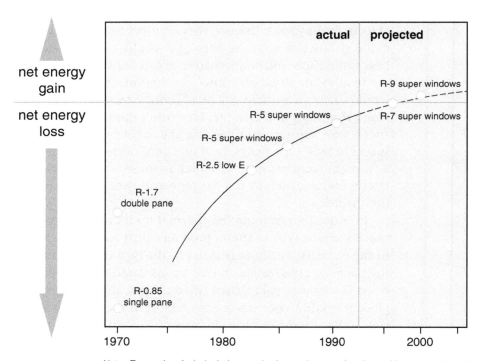

Note: Energy loss/gain includes conductive and convective thermal losses, solar gains and the effects of window emittance and lighting load reductions.

Advanced window glazings have increased windows' resistance to heat flow, or R-value.

Figure 10.10   *Effect of advanced window glazing.*

Smart Windows

Liquid crystals, suspended particle devices (SPDs), and electrochromics are being currently developed as promising window technologies—with reflective hydrides nipping closely at their heels. This new type of window uses small, light-absorbing microscopic particles known as suspended particle devices (SPDs), or light valves, to make it go from clear to dark in a matter of seconds.

The working principle is as follows. In an SPD window, millions of these SPDs are placed between two panels of glass or plastic, which are coated with a transparent conductive material. When electricity contacts the SPDs via the conductive coating, they line up in a straight line and allow light to flow through. Once the flow of electricity is stopped they move back into a random pattern and block light. When the amount of voltage is decreased, the window darkens until it becomes completely dark when all electricity flow stops.

Table 10.6 shows the cost effectiveness of replacing old windows with new and improved windows. The costs are calculated using a computer program called RESFEN developed by the U.S. Department of Energy.

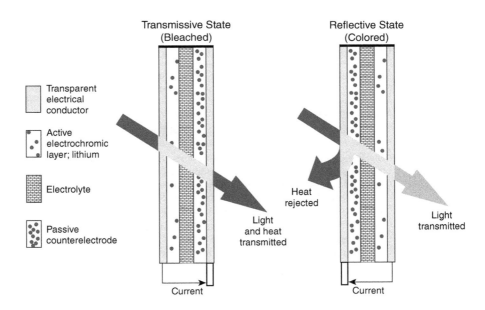

Figure 10.11 *A schematic showing the operation of a smart window.*

Table 10.5 *Cost Effectiveness of Using Improved Windows*

COST-EFFECTIVENESS EXAMPLE			
PERFORMANCE	BASE MODEL	RECOMMENDED LEVEL	BEST AVAILABLE
Window Description	Double-paned, clear glass, aluminum frame	Double-paned, low-e coating, wood or vinyl frame	Triple-paned, tinted, two spectrally selective low-e coatings, krypton-filled, wood or vinyl frame
SHGC[a]	0.61	0.55	0.20
U-factor[b]	0.87	0.40	0.15
Annual Heating Energy Use	547 therms	429 therms	426 therms
Annual Cooling Energy Use	1,134 kWh	1,103 kWh	588 kWh
Annual Energy Cost	$290	$240	$210
Lifetime Energy Cost[c]	$4,700	$3,900	$3,400
Lifetime Energy Cost Savings	—	$800	$1,300

[a] SHGC, or Solar Heat Gain Coefficient, is a measure of the solar radiation admitted through a window. SHGC ranges between 0 and 1; the lower the number, the lower the transmission of solar heat. SHGC has replaced the shading coefficient (SC) as the standard indicator of a window's shading ability. SHGC is approximately equal to the SC multiplied by 0.87.

[b] The U-factor is a measure of the rate of heat flow through a window. The U-factor is the inverse of the R-value, or resistance, the common measure of insulation.

[c] Lifetime energy cost savings is the sum of the discounted value of annual energy cost savings, based on average usage and an assumed window life of 25 years. Future energy price trends and a discount rate of 3.4 percent are based on federal guidelines (effective from April 2000 to March 2001). Assumed electricity price: $0.06/kWh, the federal average electricity price in the U.S. Assumed gas price: $0.40/therm, the federal average gas price in the U.S.

Cost-Effectiveness Assumptions: The model shown above is the result of a simulation using a residential windows modeling program called RESFEN. Calculations are based on a prototype house: 1,540 sq. ft., two stories, a standard efficiency gas furnace and central air conditioner, and window area covering 15 percent of the exterior wall surface area.

Illustration 10.1

A house in State College, PA has 380 ft² of windows (R = 1.1), 2,750 ft² of walls, and 1,920 ft² of roof (R = 30). The composite R-value of the walls is 19. Calculate the heating requirement for the house for the heating season. What is the percentage of heat that is lost through the windows?

Solution: Heat loss in a heating season is given by:

$$Heat\ loss = \frac{Area \times HDD \times 24}{R\text{-}Value}$$

Heat loss through windows =

$$\frac{380\,ft^2 \times 6,000°\cancel{K}\,days \times 24\,\cancel{h}\ per\ day}{1.1\frac{ft^2 \cdot °\cancel{K} \cdot \cancel{h}}{BTUs}} = 49,745,455\ BTUs$$

Heat loss through walls =

$$\frac{2,750\,ft^2 \times 6,000°\cancel{K}\,days \times 24\,\cancel{h}\ per\ day}{19\frac{ft^2 \cdot °\cancel{K} \cdot \cancel{h}}{BTUs}} = 20,842,105\ BTUs$$

Heat loss through the roof =

$$\frac{1,920\,ft^2 \times 6,000°\cancel{K}\,days \times 24\,\cancel{h}\ per\ day}{30\frac{ft^2 \cdot °\cancel{K} \cdot \cancel{h}}{BTUs}} = 9,216,000\ BTUs$$

Total heat loss = 79,803,560 BTUs

Percentage of heat loss through the windows = $\dfrac{49.74\ MMBTUs}{79.8\ MMBTUs} \times 100 = 62.3\%$

Illustration 10.2

Windows in the house described in Illustration 10.1 are upgraded at a cost of $1,550. The upgraded windows have an R-value of 4.0.

    a. What is the percent savings in the energy and heating bill if the energy cost is 11.15/MMBTUs?

    b. What is the payback period for this modification?

Solution:

a. New heat loss for the same window size with the new R-value is:

$$\frac{380\,ft^2 \times 6,000°K\,days \times 24\,K\,per\,day}{4.0\,\dfrac{ft^2 \cdot °K \cdot K}{BTUs}} = 13,680,000\,BTUs$$

Annual savings in the energy =

        49.745 MMBTUs −13.680 MMBTUs = 36.06 MMBTUs

The percent savings is 36.06/79.84 × 100 = 45.1%

The old heating bill would be 79.803 MMBTU $\dfrac{\$11.15}{MMBTU}$ = $889.80

The new heating bill would be 49.745 MMBTU $\dfrac{\$11.15}{MMBTU}$ = $554.65

The monetary savings = $335.14 per year

b. The payback period = $\dfrac{\text{Additional investment}}{\text{Savings per year}}$ = $\dfrac{\$1,550.00}{\$325.14}$ = 4.62 years

Sources

http://www.eere.energy.gov/buildings/info/homes/buyingwindows.html
http://windows.lbl.gov/pub/selectingwindows/window.pdf
http://www.energystar.gov/index.cfm?c=windows_doors.pr_tips_windows
http://www.nfrc.org/
http://www.energystar.gov/index.cfm?c=windows_doors.pr_anat_window#glazings
http://eetd.lbl.gov/success/window.html
Warner, J. (July/August 1995). Selecting windows for energy efficiency. *Home Energy Magazine* Online, *http://hem.dis.anl.gov/eehem/95/950708.html*
http://www.efficientwindows.org/

questions

1. What are the main factors that are important in selecting a window?

2. Explain briefly what technological advances account for increase in the performance of windows.

3. What are gas-filled windows, and how do they perform better than regular windows?

4. Is it true that a person in Alaska requires a high SHGC for windows?

5. If the thickness of the window is doubled, what will be the new percentage of heat loss?

6. What will be the heat loss from a window if there is a vacuum between two window panes?

multiple choice questions

1. Air space between the window panes does have an effect on window R-values.
   a. True
   b. False

2. In wintertime when the drapes are open for energy efficiency, the most important type of radiation that is allowed is
   a. UV radiation
   b. IR radiation
   c. Visible light
   d. X rays

3. The R-value for a window is the reciprocal of
   a. U-Value or U-factor
   b. H-Value
   c. GWP
   d. Efficiency Factor

4. Which of the following is not a consideration when picking a window?
   a. SHGC
   b. VT
   c. AL
   d. Cp

5. Window coatings that only let through certain wavelengths are called
   a. Low-e coatings
   b. Spectrally selective coatings
   c. Filter films
   d. Absorption films

6. These coatings on windows come in soft and hard coats.
   a. Spectrally selective
   b. Inert gas
   c. Low-e glazings
   d. Heat-absorbing coatings

7. Inert gas between two window panes reduces
   a. Convective heat loss between the panes
   b. Radiative heat loss
   c. Helium
   d. Air

8. Windows can be designed not only to prevent heat loss but also for heat gain.
   a. True
   b. False

9. The fraction of infrared radiation admitted by a window is characterized by
   a. SHGC
   b. VT
   c. AL
   d. Cp

10. Spacer materials used in a window reduce
    a. Conduction losses
    b. Convective losses
    c. Radiation losses

problems for practice

1. A glazing material costs about $1/ft² and improves the insulation (R-value) by 10%. What is the payback period of adding the glazing if the original window costs $10/ft² and heat loss per sq. ft from the window is 0.5 MMBTU/year? Assume the cost of energy as $10/MMBTU.

2. A house in State College, PA consists of the following: 12 single-pane windows (each 6 ft by 3 ft with an R-value of 1). Calculate the total number of BTUs lost for one season through these 12 windows. HDD for State College are 6,000.

$$Seasonal\ heat\ loss\ (BTUs) = \frac{Area(ft^2) \times HDD(°F\ days) \times 24\frac{h}{day}}{R\text{-}Value\left(\frac{ft^2\ °F\ hr}{BTU}\right)}$$

3. Heat loss through a window (R-value = 2) is 10 MMBTU/year. Calculate the payback period if each of the following gas fillings is used. Assume the energy cost to be $10/MMBTU.

GAS FILL	ADDITIONAL COST FOR THE FILLING	EFFECTIVE R	PAYBACK PERIOD (YEARS)
Argon	$20	7	
Krypton	$45	12	
Xenon	$75	20	

Windows Puzzle

Across

4. One would like to have this as high as possible to reduce artificial light inside

5. This is a better insulating material than glass

6. The low value of this coating helps reflect more light

9. This factor is reciprocal of R-value

14. Low-e coatings generally reduce heat loss through this mechanism

15. These occupy about 15 percent of the area of a typical house

17. When windows are filled with this gas, it will reduce heat losses

Down

1. A coating that modifies the property of the window glass

2. Glass layers

3. This radiation degrades the fabrics

7. Windows should maximize this way of lighting

8. An important factor in selecting windows

10. A material that is used for window frames

11. These reduce the conductivity of the metallic frames

12. This gas is much better for filling than argon and krypton

13. A gas that has low conductivity and low viscosity

16. The fraction of heat energy that is allowed through a window

Explanation of Selected Terms Related to Energy and Environmental Protection

Acid Deposition: A complex chemical and atmospheric phenomenon that occurs when emissions of sulfur and nitrogen compounds and other substances are transformed by chemical processes in the atmosphere, often far from the original sources, and then deposited on Earth in either wet or dry form. The wet forms, popularly called "acid rain," can fall to earth as rain, snow, or fog. The dry forms are acidic gases or particulates.

AFUE: Annual Fuel Utilization Efficiency. A measure of a furnace's heating efficiency. The higher the AFUE, the more efficient the furnace is. The government's established minimum AFUE rating for furnaces is 78 percent.

Air Handler: The portion of the central air conditioning/heat pump system that moves heated/cooled air throughout a home's ductwork. In some systems a furnace handles this function.

Air Leakage Rating: A measure of the rate of infiltration around a window or a skylight. It is expressed in units of cfm/ft^2 of window area. The lower a window's air leakage rating, the greater is its air tightness.

Air Source Heat Pump: A heating-cooling unit that transfers heat in either direction between the air outside a home and the indoors.

Air-leakage: The amount of air leaking in and out of a building through cracks in walls, windows, and doors.

Annual Fuel Utilization Efficiency (AFUE): AFUE is a measure of a furnace's heating efficiency expressed as a percent. The higher the AFUE, the more efficient the product is. The government's established minimum rating for furnaces is 78 percent.

Balance Point: An outdoor temperature, usually between 30°F and 45°F, at which a heat pump's output exactly equals the heating needs of the home. Below the balance point, supplementary electric resistance heat is needed to maintain indoor comfort.

Ballast: Transforms and controls electrical power to the light. A device used with an electric-discharge lamp (for instance, fluorescent lamps) to obtain the necessary circuit conditions for starting and operating.

British Thermal Unit (BTU): Unit of heat energy equal to the amount of heat required to raise the temperature of one pound of water by one degree Fahrenheit at sea level.

Buffer: A solution or liquid whose chemical makeup is such that it minimizes changes in pH when acids or bases are added to it.

Carbon Monoxide (CO): A colorless, odorless, poisonous gas produced by incomplete fossil fuel combustion.

Coefficient of Performance (COP): A ratio calculated by dividing the total heating capacity provided by the heat pump, including circulating fan heat but excluding supplementary resistance heat (BTUs per hour), by the total electrical input (watts) $\times$ 3.412.

Compact Fluorescent Light Bulbs (CFLs): Small-diameter fluorescent lamps, twisted and folded for compactness. Some feature a round adapter, allowing them to screw into common electrical sockets and making them ideal replacements for incandescent bulbs. Newer fixtures allow CFLs to be plugged in directly without the round adapters. CFLs are designated as T-12 or T-8; T-12's are 1.5 inch in diameter while T-8's are 1 inch in diameter. Lamps that are thinner block less of their own light.

Compressor: The compressor is the "engine" that drives the condensing unit. The condensing unit serves as a pump that compresses the gas in the high pressure (condensing) side of the cooling cycle and causes the refrigerant (Freon) to circulate.

Condensation: The deposit of water vapor from the air on any cold surface whose temperature is below the dew point, such as a cold window glass or frames that are exposed to humid indoor air.

Condensing Unit: It is also known as the "outdoor unit." The condensing unit pumps vaporized refrigerant from the air handler (indoor unit), compresses it, liquifies it, and returns it. It contains the compressor coil, an outdoor fan motor, refrigerant control valves, and other necessary controls.

Conduction: Heat transfer through a solid material by contact of one molecule to the next. Heat flows from a higher-temperature area to a lower-temperature one.

Convection: A heat transfer process involving motion in a fluid (such as air) caused by the difference in density of the fluid and the action of gravity. Convection affects heat transfer from the glass surface to room air, and between two panes of glass.

Dobson Unit (DU): Units of ozone level measurement. If, for example, 100 DU of ozone were brought to the Earth's surface they would form a layer one millimeter thick. Ozone levels vary geographically, even in the absence of ozone depletion.

EER: Energy Efficiency Ratio ratings are determined by dividing the cooling output of the ground or water source heat pump (in BTU/hour) by the power input (in watts)

EF: Energy Factor is used to rate the energy efficiency of storage-type hot water heaters, clothes washers, and dishwashers

Electricity: Electrical energy measured in kilowatt hours (kWh).

Electronic Ballast: Electronic ballasts convert power to light more efficiently than older magnetic ballasts and provide the same amount of light while reducing energy use up to 25 percent. Electronic ballasts represent a major step forward in energy-efficient lighting, offering savings in cost, energy, and pollution. Typically, electronic ballasts require bulbs designed for use with this type of ballast.

Emissivity: The relative ability of a surface to radiate heat.

Energy Efficiency Ratio (EER): A ratio calculated by dividing the cooling capacity in BTUs per hour (BTU/h) by the power input in watts at a given set of rating conditions, expressed in BTU/h per watt. (*See* Seasonal Energy Efficiency Ratio.)

Energy: Broadly defined as the capability to do work. In the electric power industry, energy is more narrowly defined as electricity supplied over time, expressed in kilowatts.

Evaporator Coil: The portion of a heat pump or central air conditioning system that is located in the home and functions as the heat transfer point for warming or cooling indoor air.

Fixtures: Devices that contain the bulbs and (if necessary) the ballasts. Fixtures come in hundreds of shapes and sizes depending on their application. Some are designed to give architectural appeal and provide ambiance, and some are specially designed to minimize glare on computer screens and other working surfaces.

Flue Gas: The air coming out of a chimney after combustion in the burner it is venting. It can include nitrogen oxides, carbon oxides, water vapor, sulfur oxides, particles and many chemical pollutants.

Fluorescent Lamps: Low-pressure-mercury electric discharge lamps in which a fluorescing coating transforms some of the ultraviolet energy generated into light. Although fluorescent lamps are often associated with a harsh white light, lamps that simulate natural daylight and incandescent light are now available. These are often rated by the color temperature and color rendering index. *See* Compact Fluorescent Light Bulbs.

Foot-candle (fc): A unit of illumination or light falling onto a surface. One foot-candle is equal to one lumen per square foot.

Fuel Oil: The default units are gallons.

Gas Filled Units: Insulating glass units with a gas other than air in the air space; to decrease the unit's thermal conductivity U-value.

Gigawatt: A unit of electric power equal to one billion watts, or one thousand megawatts.

Glazing: The glass or plastic panes in a window or skylight.

Global Warming Potential (GWP): The ratio of the warming caused by a substance to the warming caused by a similar mass of carbon dioxide. CFC-12, for example, has a GWP of 8,500, while water has a GWP of zero.

Global Warming: An increase in the near surface temperature of the Earth. Global warming has occurred in the distant past as the result of natural influences, but the term is most often used to refer to the warming predicted to occur as a result of increased emissions of greenhouse gases. Scientists generally agree that the Earth's surface has warmed by about 1° Fahrenheit in the past 140 years. The Intergovernmental Panel on Climate Change (IPCC) recently concluded that increased concentrations of greenhouse gases are causing an increase in the Earth's surface temperature and that increased concentrations of sulfate aerosols have led to relative cooling in some regions, generally over and downwind of heavily industrialized areas.

Greenhouse Effect: The warming of the Earth's atmosphere attributed to a buildup of carbon dioxide or other gases; some scientists think that this buildup allows the sun's rays to heat the Earth, while making the infrared radiation atmosphere opaque to infrared radiation, thereby preventing a counterbalancing loss of heat.

Greenhouse Gas: A gas, such as carbon dioxide or methane, which contributes to potential climate change.

Halogen Lamp: A short name for the tungsten-halogen lamp. Halogen lamps are high pressure incandescent lamps containing halogen gases such as iodine or bromine, which allow the filaments to be operated at higher temperatures and higher efficacies. A higher temperature chemical reaction involving tungsten and the halogen gas recycles evaporated particles of tungsten back onto the filament surface.

Heat Pump: An electric device with both heating and cooling capabilities. It extracts heat from one medium at a lower (the heat source) temperature and transfers it to another at a higher temperature (the heat sink), thereby cooling the first and warming the second.

Heat Source: A body of gas, liquid or solid, from which heat is collected. In an air-to-air heat pump the air outside the home is used as the heat source during the heating cycle.

Heating Degree Day: Term used by heating and cooling engineers to relate the typical climate conditions of different areas to the amount of energy needed to heat and cool a building. The base temperature is 65° Fahrenheit. A heating degree day is counted for each degree below 65° reached by the average daily outside temperatures in the winter. For example, if on a given winter day, the daily average temperature outdoors is 30°, then there are 35° below the base temperature of 65°. Thus, there are 35 heating degree days for that day.

Heating Seasonal Performance Factor (HSPF): HSPF is typically used with heat pumps. The higher the HSPF rating, the more efficient a heat pump is at heating your building.

High Intensity Discharge Lamp (HID Lamp): An electric discharge lamp, including groups of lamps known as mercury, metal halide, and high pressure sodium. The major portion of the light is produced by radiation of metal halides and their products of dissociation— possibly in combination with metallic vapors such as mercury.

Incandescent Lamp: A lamp in which light is produced by a filament heated to incandescence by an electric current. Typical light bulbs are incandescent lamps. These bulbs are typically very inefficient, converting only about 5–10 percent of the energy to light—the rest is transformed into heat.

Infiltration: The inadvertent flow of air into a building through breaks in the exterior surfaces of the building. It can occur through joints and cracks around window and skylight frames, sash, and glazings.

Kerosene: A liquid fuel used to heat homes that is measured in gallons.

Kilowatt (KW): A measurement of electric power equal to 1,000 watts. Electric power capacity of 1 kW is sufficient to power ten 100-watt light bulbs.

Kilowatt Hour (kWh): A measurement of energy equal to the energy produced by a 1,000-watt plant in one hour.

Light (photo) Sensors: Are used to integrate the building's electric lighting system with its natural daylighting system, so that the lights go on only when daylighting is insufficient.

Low-Conductance Spacers: An assembly of materials designed to reduce heat transfer at the edge of an insulating window. Spacers are placed between the panes of glass in a double- or triple-glazed window.

Low-Emissivity (low-e) Coating: a microscopically thin, virtually invisible, metal or metallic oxide layer deposited on a window or skylight glazing surface to reduce the U-factor or solar heat gain coefficient by suppressing radiative heat flow through the window or skylight.

Lumen Per Watt (lpw): A measure of the efficiency, or, more properly, "efficacy" of a light source. Efficacy is easily calculated by taking the lumen output of a lamp and dividing by the lamp watts. For example, a 100-watt light source producing 1,750 lumens has an efficacy of 17.5 lumens per watt.

Luminous Efficacy: The light output of a light source divided by the total power input to that source. It is expressed in lumens per watt.

Megawatt (MW): A unit of electric power equal to 1 million watts, or 1,000 kilowatts.

Montreal Protocol: Treaty, signed in 1987, governs stratospheric ozone protection and research, and the production and use of ozone-depleting substances. It provides for the end of production of ozone-depleting substances such as CFCS. Under the Protocol, various research groups continue to assess the ozone layer. The Multilateral Fund provides resources to developing nations to promote the transition to ozone-safe technologies.

Motion Sensors: are used so that the lights go on when someone is in the space. These save energy by not depending on people to turn lights off as they leave a room. They also provide convenience and security when used for outdoor lighting, while minimizing the use of the lighting.

Natural Gas: A gaseous fuel that is formed naturally in the ground. The commonly used units are hundred cubic feet (CCF). CCF is the most common units used for natural gas, but your bill may show your natural gas use in therms. Practically speaking, one CCF is equal to one therm, so you can enter either value and the calculations will work correctly. It is sometime measured in MCF (thousand cubic feet). Typically 1 MCF of natural gas has 1 million BTUs.

Nitric Oxide (NO): A gas formed by combustion under high temperature and high pressure in an internal combustion engine; it is converted by sunlight and photochemical processes in ambient air to nitrogen oxide. NO is a precursor of ground-level ozone pollution, or smog.

Outdoor Coil/Condensing Unit: The portion of a heat pump or central air conditioning system that is located outside the home and functions as a heat transfer point for collecting heat from and dispelling heat to the outside air.

Ozone (O_3): Found in two layers of the atmosphere, the stratosphere and the troposphere. In the stratosphere (the atmospheric layer 10 miles or more above the Earth's surface) ozone is a natural form of oxygen that provides a protective layer shielding the Earth from ultraviolet radiation. In the troposphere (the layer extending 10 miles from the Earth's surface), ozone is a chemical oxidant and major component of photochemical smog. It can seriously impair the respiratory system and is one of the most widespread of all the criteria pollutants for which the Clean Air Act required EPA to set standards. Ozone in the troposphere is produced through complex chemical reactions of nitrogen oxides, which are among the primary pollutants emitted by combustion sources; hydrocarbons, released into the atmosphere through the combustion, handling and processing of petroleum products; and sunlight.

Ozone Hole: A thinning break in the stratospheric ozone layer. Designation of amount of such depletion as an "ozone hole" is made when the detected amount of depletion exceeds 50 percent. Seasonal ozone holes have been observed over both the Antarctic and Arctic regions, part of Canada, and the extreme northeastern United States.

Ozone Layer: The protective layer in the atmosphere, about 10 miles above the ground, that absorbs some of the sun's ultraviolet rays, thereby reducing the amount of potentially harmful radiation that reaches the earth's surface.

Parts Per Billion (ppb)/Parts Per Million (ppm): Units commonly used to express contamination ratios, as in establishing the maximum permissible amount of a contaminant in water, land, or air.

Petroleum: Crude oil or any fraction thereof that is liquid under normal conditions of temperature and pressure. The term includes petroleum-based substances comprising a complex blend of hydrocarbons derived from crude oil through the process of separation, conversion, upgrading, and finishing, such as motor fuel, jet oil, lubricants, petroleum solvents, and used oil.

pH: An expression of the intensity of the basic or acid condition of a liquid; may range from 0 to 14, where 0 is the most acid and 7 is neutral. Natural waters usually have a pH between 6.5 and 8.5.

Photochemical Oxidants: Air pollutants formed by the action of sunlight on oxides of nitrogen and hydrocarbons.

PM-10/PM-2.5: PM 10 is a measure of particles in the atmosphere with a diameter of less than 10 or equal to a nominal 10 micrometers. PM-2.5 is a measure of smaller particles in the air. PM-10 has been the pollutant particulate level standard against which EPA has been measuring Clean Air Act compliance. On the basis of newer scientific findings, the Agency is considering regulations that will make PM-2.5 the new "standard."

Pollutant: Generally, any substance introduced into the environment that adversely affects the usefulness of a resource or the health of humans, animals, or ecosystems.

Radiation: The transfer of heat in the form of electromagnetic waves from one separate surface to another. Energy from the sun reaches the Earth by radiation, and a person's body can lose heat to a cold window or skylight surface in a similar way.

Reflectance: The ratio of reflected radiant energy to incident radiant energy.

Relative Humidity: The percentage of moisture in the air in relationship to the amount of moisture the air could hold at that given temperature. At 100 percent relative humidity, moisture condenses and falls as rain.

Refrigerant: Refrigerant is a substance that absorbs heat by changing states (evaporating) from liquid to gas. It releases heat by changing states (condensing) from gas back to liquid.

R-Value: A measure of the resistance of a material or assembly to heat flow. It is the inverse of the U-factor ($R = 1/U$) and is expressed in units of hr-ft^2-F/BTU. The higher a window's R-value, the greater is its resistance to heat flow and its insulating value.

Seasonal Energy Efficiency Ratio (SEER): SEER is a measure of cooling efficiency for central air conditioning products. The higher the SEER rating number, the more energy efficient the unit is.

Shading Coefficient (SC): A measure of the ability of a window or skylight to transmit solar heat, relative to that ability for ⅛-inch clear, double-strength, single glass. It is being phased out in favor of the solar heat gain coefficient and is approximately equal to the SHGC multiplied by 1.15. It is expressed as a number without units between 0 and 1. The lower a window's solar heat gain coefficient or shading coefficient, the less solar heat it transmits, and the greater is its shading ability.

Packaged Unit: A year-round heating and air conditioning system that has all of the components completely encased in one unit outside the home.

Solar Heat Gain Coefficient (SHGC): The fraction of solar radiation admitted through a window or skylight, both directly transmitted, absorbed, and subsequently released inward. The solar heat gain coefficient has replaced the shading coefficient as the standard indicator of a window's shading ability. It is expressed as a number between 0 and 1. The lower a window's solar heat gain coefficient, the less solar heat it transmits, and the greater its shading ability. SHGC can be expressed in terms of the glass alone or it can refer to the entire window assembly.

Spectrally Selective Coating: A coated or tinted glazing with optical properties that are transparent to some wavelengths of energy and reflective to others. Typical spectrally selective coatings are transparent to visible light and reflect short-wave and long-wave infrared radiation.

Split System: A heat pump or central air conditioning system with components located both inside and outside the home. The most common design for home use.

Sulfur Dioxide (SO$_2$): A pungent, colorless, gas formed primarily by the combustion of fossil fuels; becomes a pollutant when present in large amounts.

Super Window: A window with a very low U-factor, typically less than 0.15, achieved through the use of multiple glazings, low-e coatings, and gas fills.

Supplementary Heat: The auxiliary or emergency heat, usually electrical resistance heat, provided at temperatures below a heat pump's balance point.

Thermostat: A temperature-sensitive switch that controls your heating and cooling systems. When the indoor temperature drops below or rises above the selected temperature setting, the switch moves to the "on" position, and your heater or air conditioner runs to warm or cool.

Transmittance: The percentage of radiation that can pass through glazing. Transmittance can be defined for different types of light or energy, e.g., visible light transmittance, UV transmittance, or total solar energy transmittance.

U-factor (U-value): A measure of the rate of heat flow through a material or assembly. The reciprocal of R-value. It is expressed in units of BTU/hr-ft^2-°F. Window manufacturers and engineers commonly use the U-factor to describe the rate of non-solar heat loss or gain through a window or skylight. The lower a window's U-factor, the greater its resistance to heat flow and its insulating value.

Ultraviolet Light (UV): The invisible rays of the spectrum that are outside of the visible spectrum at its short-wavelength violet end. Ultraviolet rays are found in everyday sunlight and can cause fading of paint finishes, carpets, and fabrics.

Ultraviolet Rays: Radiation from the sun that can be useful or potentially harmful. UV rays from one part of the spectrum (UV-A) enhance plant life. UV rays from other parts of the spectrum (UV-B) can cause skin cancer or other tissue damage. The ozone layer in the atmosphere partly shields us from ultraviolet rays reaching the Earth's surface

Visible Light: The portion of the electromagnetic spectrum that produces light that the human eye detects as light. Wavelengths range from 380 to 720 nanometers.

Visible Transmittance (VT): The percentage or fraction of the visible spectrum (380 to 720 nanometers) weighted by the sensitivity of the eye that is transmitted by a window or skylight.

Volatile Organic Compound (VOC): Any organic compound that participates in atmospheric photochemical reactions except those designated by EPA as having negligible photochemical reactivity.

Weatherstripping: A strip of resilient material for covering the joint between the window sash and frame in order to reduce air leaks and prevent water from entering the structure.

Answers to Multiple Choice Questions

Chapter 1

1. c 2. b 3. b 4. d 5. b 6. b 7. a 8. d 9. a 10. a 11. c 12. a 13. a 14. d 15. b
16. a 17. a 18. a 19. c 20. a 21 b. 22. a 23. a 24. b. 25. b 26. a

Chapter 2

1. a 2. b 3. d 4. b 5. b 6. b 7. b 8. b 9. b 10. b 11. a 12. c

Chapter 3

1. a 2. b 3. c 4. c 5. b 6. b 7. b 8. b 9. b 10. d 11. a 12. c 13. b 14. d 15. d
16. b 17. c 18. a 19. b

Chapter 4

1. a 2. a 3. a 4. a 5. a 6. b 7. a 8. b 9. c 10. d 11. a 12. d 13. c 14. d 15. a
16. d 17. b 18. b 19. a 20. a 21. c 22. c 23. d 24. a 25. d 26. b 27. c

Chapter 5

1. e 2. d 3. a 4. b 5. b 6. b 7. c 8. a 9. b 10. D

Chapter 6

1. a 2. d 3. b 4. d 5. c 6. b 7. a 8. d 9. a 10. b 11. b 12. c 13. b 14. a, b
15. b 16. c 17. a 18. b 19. a 20. b 21. a 22. b 23. b 24. c 25. a 26. b 27. b
28. a 29. d

Chapter 7

1. c 2. b 3. a 4. d 5. a 6. c 7. b 8. b 9. b 10. a 11. a 12. b 13. c 14. a 15. d
16. a

Chapter 8

1. d   2. c   3. c   4. b   5. b   6. d   7. d   8. b   9. a   10. d   11. d   12. a   13. c   14. c   15. c   16. a   17. b   18. d   19. b   20. c   21. c   22. b   23. a

Chapter 9

1. d   2. a   3. b   4. a   5. b   6. a   7. d   8. b   9. c   10. a   11. c   12. a   13. a   14. b   15. b   16. a   17. a   18. a   19. a   20. a   21. a   22. b   23. b

Chapter 10

1. a   2. b   3. a   4. d   5. b   6. c   7. a   8. a   9. a   10. a

$\mathcal{A}$nswers to Numerical Problems

Chapter 1

1. 375 Cal   2. $657   3. 9 kWh   4. $ 48.00
5. Jackie 18 kWh, Stacie 108 kWh, Difference in cost $7.65
6. $29.2   7. $16.42   8. $1,460

Chapter 3

1. 2333.3 J   4. 76.9%

Chapter 5

1. 15,000 BTU   2. 36,500,000 BTU   3. 50 kWh   4. 24.4%   5. $4.0/day
6. 1.24 CCF   7. $37.5   8. $87.6   9. 146 kWh, $11.7 every month
10. 18.18%   11. $65.7   12. yes, $131.0, yes   13. yes, –$40, no

Chapter 6

1. 5%   2. $0.12   3. $0.1   4. $48.86   5. $1,172.64

Chapter 7

1. 52 degree-days
2. 29.4 degree-days
3. 60 degree-days
4. 1,240 degree-days
5. 6,138 degree-days
6. 114%
7. 15.2
8. 2.94
9. 19.76
10. 22.04
11. 22.58
12. 160 sq. ft
13. 448 sq. ft
14. 28 sq. ft, 36 sq. ft, 576 sq. ft
15. 98 BTU/hr
16. 256 BTU/hr
17. 5,498 BTU
18. 7,061 BTU/hr
19. 34,492 BTU
20. 1.39 million BTU
21. 1.58 million BTU
22. 712,800 BTU
23. 9.63 million BTU
24. 2.23 million BTU
25. 113.14 million BTU
26. 73.18 million BTU
27. 3.4 million BTU
28. 11.62 million BTU
29. 2,360 CCF
30. 1,236 gallons
31. 3,107 KWh
32. 1,289 CCF
33. 2,292 gallons
34. $11.11/MMBTU
35. $7.71/MMBTU
36. $35.17/MMBTU
37. $0.28/MMBTU
38. $15.12/MMBTU
39. $10.71/MMBTU
40. Oil is cheaper
41. 8.13 years
42. 19.2 years
43. 16.69 years
44. 22.9 years
45. 8.47 year
46. 0.33 years (so good investment)

Chapter 8

1. $T_{hot} = 360°K$ 2. COP kitchen = 18.8 COP garage = 10.9; It's a bad idea.

Chapter 9

1. 60,000 BTUs   2. 1,000 Watts   3. 50%, yes
4. Payback period, 3.7 years is less than lifetime (10 yrs). So go for it.

Chapter 10

1. 2.2 years   2. 31.1 MMBTU   3. Ar—0.28 years, Kr—0.54 years, and Xe—0.83 years

Useful Conversion Factors

1 kWh = 3,412 BTU

1 BTU = 1,055 J

1 quadrillion BTUs = 1×10^{15} BTUs

1 mile = 5,280 ft

1 cal = 4.2 Joules

1 Cal (food calorie) = 1,000 cal

1 CCF = 100 cu. ft

1 MCF = 1,000 cu. ft

1 Therm = 100,000 BTUs

1 gal of water = 8.3 lbs of water

Cp (Heat Capacity) of water = 1 BTU/lb °F

1 foot-candle = 1 lumen/sq. ft

$$°C = \frac{(°F - 32)}{1.8}$$

K = 273 + °C

FUEL TYPE	NO. OF BTU/UNIT
Kerosene (No. 1 Fuel Oil)	135,000/gallon
No. 2 Fuel Oil (home heating oil)	140,000/gallon
Electricity	3,412/kWh
Natural Gas	1,000,000/thousand cubic feet (MCF)
Propane	100,000/CCF (Therm) 91,333/gallon

State College, PA HDD is 6,000°F Days

Home heating oil is No. 2 Fuel Oil

Useful Formulae

Chapter 1

- $Power = \dfrac{Energy}{Time}$

- Power Consumption per day = Power Consumption × Hours Used per Day

Chapter 2

- Doubling time $= \dfrac{70}{\% \text{ Growth rate per year}}$

Chapter 3

- $Efficiency = \dfrac{Useful\ energy\ output}{Total\ energy\ output}$

- $\eta = \left(1 - \dfrac{T_{Cold}}{T_{Hot}}\right) \times 100\%$

  where η = Carnot Efficiency
  T_{Hot} = Temperature of combustion gases inside the engine in (Kelvin)
  T_{Cold} = Temperature of exhaust gases from the engine (Kelvin)

- Overall Efficiency = product of all the step efficiencies

Chapter 5

- $Q = m \times C_p \times (\Delta t)$

  where Q = Heat required
  m = Mass of water heated
  C_p = Heat capacity of water
  Δt = Temperature difference

- $Payback\ period = \dfrac{Additional\ investment}{Savings\ per\ year}$

Chapter 6

- $LSG = \dfrac{SHGC}{VT}$

  where LSG = Light-to-solar ratio
  $SHGC$ = Solar heat gain coefficient
  VT = Visual transmittance

- Life Cycle Costs = Cost to buy + Cost to maintain it (if any maintenance is required) + Cost of energy to run it for its life + Replacement cost – Any salvage value

- Efficacy of light bulb $= \dfrac{\text{lumens}}{\text{watts}}$

Chapter 7

- $Heat\ Loss\left(\dfrac{BTUs}{h}\right) = \dfrac{Area\,(ft^{2)}) \times Temperature\ Difference\ (°F)}{R\text{-}Value\left(\dfrac{ft^2\ °F\ h}{BTU}\right)}$

- $HDD = T_{base} - T_a$

  where HDD = Heating degree days
  T_{base} = Base temperature or inside temperature, usually 65°F
  T_a = Average Outside temperature = $T_{max} + T_{min}/2$

- $Heat\ Loss\ in\ a\ Season = \dfrac{Area}{R\text{-}value} \times 24 \times HDD$

  where HDD = Heating degree days/season

- $Actual\ Energy\ Cost = \dfrac{FuelCost\left(\dfrac{\$}{Unit\ of\ Fuel}\right)}{Heating\,Value\left(\dfrac{MMBTUs}{Unit\ of\ Fuel}\right) \times Efficiency}$

- $Years\ to\ Payback = \dfrac{C_i \times R_1 \times R_2 \times E}{C_e \times [R_2 - R_1] \times HDD \times 24}$

  where C_i = Cost of insulation in \$/square feet
  C_e = Cost of energy, expressed in \$/BTUs
  E = Efficiency of the heating system
  R_1 = Initial R-value of section
  R_2 = Final R-value of section
  $(R_2 - R_1)$ = R-value of additional insulation being considered
  HDD = Heating degree days/year
  24 = Multiplier used to convert HDD to heating hours (24 hours/day)

Chapter 8

■ $COP = \left(\dfrac{T_{Hot}}{T_{Hot} - T_{Cold}} \right)$

> where COP = Coefficient of Performance
> T_{Hot} = Temperature of heat input at high temperature (Kelvin)
> T_{Cold} = Temperature of heat output at low temperature (Kelvin)

■ $Collector's\ Efficiency = \left(\dfrac{Useful\ energy\ delivered}{Insolation\ on\ collector} \right) \times 100\%$

Chapter 9

■ $Absolute\ humidity = \dfrac{Mass\ of\ water\ vapor\ (lb)}{Mass\ of\ dry\ air\ (lb)}$

■ $Relative\ humidity_{At\ a\ given\ temperature}$

$= \left(\dfrac{Amount\ of\ water\ vapor\ (lb)}{Maximum\ amount\ of\ water\ vapor\ air\ can\ hold} \right) \times 100_{At\ that\ temperature}$

■ $Energy\ Efficiency\ Ratio,\ EER = \dfrac{BTUs/h\ removed}{Watts\ of\ power\ put\ in}$

Chapter 10

■ $U\text{-}factor = \dfrac{1}{R\text{-}value}$

$\mathscr{S}$olutions to Selected Problems

Chapter 1

Problem 1

STEP 1: Convert the amount of time Michael jogs and walks from minutes to hours.

Jogging

$$15m \times \frac{1\,hr}{60m} = .25\,hr$$

Walking

$$45m \times \frac{1\,hr}{60m} = .75\,hr$$

Step 2: Calculate how many calories Michael burns by multiplying the amount of time he jogs by how many calories he burns per hour by jogging. Also, multiply the amount of time he walks by how many calories he burns per hour by walking.

Jogging

$$\frac{600\,Cal}{hr} \times .25\,hr = 150\,Cal$$

Walking

$$\frac{300\,Cal}{hr} \times .75\,hr = 225\,Cal$$

STEP 3: Add the amount of calories burned from jogging and walking to calculate the total amount of calories burned during Michael's exercise.

$$150\,Cal + 225\,Cal = 375\,Cal$$

Problem 3

STEP 1: Convert the amount of time Michael uses his toaster from minutes to hours.

$$15m \times \frac{1\,hr}{60m} = .25\,hr$$

STEP 2: Convert the toaster power consumption from watts to kilowatts.

$$1200W \times \frac{1kW}{1000W} = 1.2kW$$

STEP 3: Calculate the monthly energy consumption by multiplying the power consumption by the number of hours it operates per day and by how many days per month.

$$1.2\,kW \times \frac{.25\,hr}{day} \times \frac{30\,days}{month} = \frac{9\,kWh}{month}$$

Problem 5

STEP 1: Convert the computer power consumption from watts to kilowatts.

$$150W \times \frac{1\,kW}{1000W} = .15\,kW$$

STEP 2: Calculate the monthly energy consumption for both Jackie and Stacie. Multiply the power consumption by the number of hours the computer operates per day and by how many days per month.

Jackie	Stacie
$.15\,kW \times \dfrac{4\,hr}{day} \times \dfrac{30\,days}{month} = \dfrac{18\,kWh}{month}$	$.15\,kW \times \dfrac{24\,hr}{day} \times \dfrac{30\,days}{month} = \dfrac{108\,kWh}{month}$

STEP 3: Calculate the monthly operating cost for both Jackie and Stacie. Multiply the monthly energy consumption by the cost of electrical energy per kilowatt-hour.

Jackie	Stacie
$\dfrac{18kWh}{month} \times \dfrac{\$.085}{kWh} = \dfrac{\$1.53}{month}$	$\dfrac{108\,kWh}{month} \times \dfrac{\$.085}{kWh} = \dfrac{\$9.18}{month}$

STEP 4: Subtract Stacie's monthly operating cost from Jackie's monthly operating cost to calculate the difference in cost to operate the computers.

$$\frac{\$9.18}{month} - \frac{\$1.53}{month} = \frac{\$7.65}{month}$$

Problem 7

STEP 1: Convert the power consumption for both light bulbs from watts to kilowatts.

Incandescent light bulb	Fluorescent light bulb
$100W \times \dfrac{1\,kW}{1,000W} = .1\,kW$	$25W \times \dfrac{1\,kW}{1,000W} = .025\,kW$

STEP 2: Calculate the annual energy consumption for both the incandescent and fluorescent light bulb. Multiply the power consumption by the number of hours the light bulb operates per day and by how many days per year.

Incandescent light bulb

$$.1\,kW \times \frac{6\,hr}{day} \times \frac{365\,days}{yr} = \frac{219\,kWh}{yr}$$

Fluorescent light bulb

$$.025\,kW \times \frac{6\,hr}{day} \times \frac{365\,days}{yr} = \frac{54.75\,kWh}{yr}$$

STEP 3: Calculate the annual operating cost for both the incandescent and fluorescent light bulb. Multiply the annual energy consumption by the cost of electrical energy per kilowatt-hour.

Incandescent light bulb

$$\frac{219\,kWh}{yr} \times \frac{\$.10}{kWh} = \frac{\$21.90}{yr}$$

Fluorescent light bulb

$$\frac{54.75\,kWh}{yr} \times \frac{\$.10}{kWh} = \frac{\$5.48}{yr}$$

STEP 4: Subtract the annual operating cost of the fluorescent light bulb from the annual operating cost of the incandescent light bulb to calculate the cost savings of the fluorescent light bulb.

$$\frac{\$21.90}{yr} - \frac{\$5.48}{yr} = \frac{\$16.42}{yr}$$

Chapter 5

Problem 1

STEP 1: Identify the equation to calculate heat required, in which Q is heat required, m is water mass, Cp is the heat capacity of water (1 BTU/lb °F) and ΔT is temperature difference.

$$Q = m \times C_p \times \Delta T$$

STEP 2: Substitute the known variables and solve for heat required.

$$Q = 200\,lbs \times \frac{1\,BTU}{lb\,°F} \times (130°F - 55°F)$$

$$Q = 15,000\,BTU$$

Problem 3

STEP 1: Calculate the heat required by substituting the known variables.

$$Q = \frac{250\,gal}{day} \times \frac{8.3\,lb}{gal} \times \frac{1\,BTU}{lb\,°F} \times (140°F - 58°F)$$

$$Q = \frac{170,150\,BTU}{day}$$

STEP 2: Convert the heat required from BTUs to kilowatt-hours.

$$\frac{170,150\,BTU}{day} \times \frac{1\,kWh}{3,412\,BTU} = \frac{49.87\,kWh}{day}$$

Problem 5

STEP 1: Calculate the daily operating cost by multiplying the daily energy consumption from problem 3 by the cost of electrical energy per kilowatt-hour.

$$\frac{49.87 \, kWh}{day} \times \frac{\$.08}{kWh} = \frac{\$3.99}{day}$$

Problem 7

STEP 1: Calculate the monthly cost by multiplying the monthly heat requirement by the cost of natural gas. Assume that 1 CCF of natural gas costs $1.00.

$$\frac{37.5 \, CCF}{month} \times \frac{\$1.00}{1 \, CCF} = \frac{\$37.5}{month}$$

Problem 9

STEP 1: Calculate the heat required if the temperature of the water heater were reduced to 120°F.

$$Q = \frac{200 \, gal}{day} \times \frac{8.3 \, lb}{gal} \times \frac{1 \, BTU}{lb \, °F} \times (120°F - 55°F)$$

$$Q = \frac{107,900 \, BTU}{day}$$

STEP 2: Calculate the monthly BTU requirement by multiplying the daily requirement by how many days per month. Then, convert the monthly heat requirement from BTUs to kilowatt-hours.

$$\frac{107,900 \, BTU}{day} \times \frac{30 \, days}{month} = \frac{3,237,000 \, BTU}{month}$$

$$\frac{3,237,000 \, BTU}{month} \times \frac{1 \, kWh}{3,412 \, BTU} = \frac{948.71 \, kWh}{month}$$

STEP 3: Calculate the monthly energy savings by subtracting the heat required for the higher temperature from that of the lower temperature.

$$\frac{1,094.67 \, kWh}{month} - \frac{948.71 \, kWh}{month} = \frac{145.96 \, kWh}{month}$$

STEP 4: Calculate the monthly cost savings by multiplying the monthly energy savings by the cost of electricity.

$$\frac{145.96 \, kWh}{month} \times \frac{\$.08}{kWh} = \frac{\$11.67}{month}$$

Problem 11

STEP 1: Convert the refrigerator power consumption from watts to kilowatts.

$$150W \times \frac{1\,kW}{1,000W} = .15\,kW$$

STEP 2: Calculate the annual energy consumption by multiplying the power consumption by the number of hours it operates per day and by how many days per year.

$$.15\,kW \times \frac{20\,hrs}{day} \times \frac{365\,days}{yr} = \frac{1,095\,kWh}{yr}$$

STEP 3: Calculate the annual operating cost by multiplying the annual energy consumption by the cost of electrical energy per kilowatt-hour.

$$\frac{1,095\,kWh}{yr} \times \frac{\$.06}{kWh} = \frac{\$65.70}{yr}$$

Problem 13

STEP 1: Convert the power consumption for both air conditioners from watts to kilowatts.

Most Efficient Model

$$250W \times \frac{1\,kW}{1,000W} = .25\,kW$$

Least Efficient Model

$$275W \times \frac{1\,kW}{1,000W} = .275\,kW$$

STEP 2: Calculate the annual energy consumption for both the most efficient and least efficient air conditioner. Multiply the power consumption by the number of hours the air conditioner operates annually.

Most Efficient Model

$$.25\,kW \times \frac{4,000\,hr}{yr} = \frac{1,000\,kWh}{yr}$$

Least Efficient Model

$$.275\,kW \times \frac{4,000\,hr}{yr} = \frac{1,100\,kWh}{yr}$$

STEP 3: Calculate the annual operating cost for both air conditioners. Multiply the annual energy consumption by the cost of electricity per kilowatt-hour.

Most Efficient Model

$$\frac{1,000\,kWh}{yr} \times \frac{\$.06}{kWh} = \frac{\$60}{yr}$$

Least Efficient Model

$$\frac{1,100\,kWh}{yr} \times \frac{\$.06}{kWh} = \frac{\$66}{yr}$$

STEP 4: Calculate the cost savings with the most efficient air conditioner. First, calculate the annual cost savings by subtracting the annual operating cost of the most efficient model from that of the least efficient model. Next, calculate the lifetime cost savings by multiplying the annual cost savings by the expected lifetime of operation.

$$\frac{\$66}{yr} - \frac{\$60}{yr} = \frac{\$6}{yr}$$

$$\frac{\$6}{yr} \times 10 \ yrs = \$60$$

ANSWER: Yes, saving money with the most efficient model.

STEP 5: Since the most efficient model costs $100 more than the least efficient model, subtract the excess cost from the cost savings to calculate the overall savings.

$$\$60 - \$100 = -\$40$$

ANSWER: No, losing money by purchasing the most efficient model.

Chapter 6

Problem 1

STEP 1: Determine the efficiency equation.

$$Efficiency = \frac{Useful \ Output}{Total \ Output} \times 100\%$$

STEP 2: Determine the value of useful output and the value for total output. The radiant energy the light bulb produces is considered useful output because we expect light from the light bulb, not heat.

$$Radiant \ Energy \rightarrow Useful \ Output = 3W$$

$$Total \ Output = 60W$$

STEP 3: Input the values for useful output and total output into the efficiency equation to calculate the efficiency of the light bulb.

$$Efficiency = \frac{3W}{60W} \times 100\% = 5\%$$

Problem 3

STEP 1: Convert the power consumption for both light bulbs from watts to kilowatts.

Incandescent light bulb	**Fluorescent light bulb**
$100W \times \dfrac{1\,kW}{1,000W} = .1\,kW$	$15W \times \dfrac{1\,kW}{1,000W} = .015\,kW$

STEP 2: Calculate the energy consumption for both the incandescent and fluorescent light bulb. Multiply the power consumption by the number of hours the light bulb operated.

Incandescent light bulb	**Fluorescent light bulb**
$.1\,kW \times 12\,h = 1.2\,kWh$	$.015 \times 12\,h = .18\,kWh$

STEP 3: Calculate the operating cost for both the incandescent and fluorescent light bulb. Multiply the energy consumption by the cost of electrical energy per kilowatt-hour.

Incandescent light bulb	**Fluorescent light bulb**
$1.2\,kWh \times \dfrac{\$.10}{kWh} = \$.12$	$.18\,kWh \times \dfrac{\$.10}{kWh} = \$.02$

STEP 4: Subtract the operating cost of the fluorescent light bulb from the operating cost of the incandescent light bulb to calculate the cost savings of the fluorescent light bulb.

$$\$.12 - \$.02 = \$.10$$

Problem 4

STEP 1: Calculate the purchasing cost for both the incandescent and fluorescent light bulb. Since the CFL light bulb life expectancy is 8 times greater than that of the incandescent light bulb, 8 incandescent light bulbs must be purchased to operate for the same amount of time as 1 CFL light bulb.

Incandescent light bulb	**CFL light bulb**
Life expectancy = 1,000 h	Life expectancy = 8,000 h
$\dfrac{8,000\,h}{1,000\,h} = $ light bulbs	
Cost = 8 light bulbs $\times$ \$.50 = \$4.00	Cost = 1 light bulb $\times$ \$7.50 = \$7.50

STEP 2: Convert the power consumption for both light bulbs from watts to kilowatts.

Incandescent light bulb	**CFL light bulb**
$100W \times \dfrac{1\,kW}{1,000W} = .1\,kW$	$23W \times \dfrac{1\,kW}{1,000W} = .023\,kW$

STEP 3: Calculate the energy consumption for both the incandescent and fluorescent light bulb. Multiply the power consumption by the number of hours the light bulb operated.

Incandescent light bulb	**CFL light bulb**
$.1\,kW \times 8,000\,h = 800\,kWh$	$.023\,kW \times 8,000\,h = 184\,kWh$

STEP 4: Calculate the operating cost for both the incandescent and fluorescent light bulb. Multiply the energy consumption by the cost of electrical energy per kilowatt-hour.

Incandescent light bulb	*CFL light bulb*
$800\,kWh \times \dfrac{\$.085}{kWh} = \$68$	$184\,kWh \times \dfrac{\$.085}{kWh} = \$15.64$

STEP 5: Calculate the total cost for both light bulbs by adding the purchasing cost and operating cost.

Incandescent light bulb	*CFL light bulb*
$\$4.00 + \$68 = \$72$	$\$7.50 + \$15.64 = \$23.14$

STEP 6: Subtract the total cost of the fluorescent light bulb from the total cost of the incandescent light bulb to calculate the cost savings of the fluorescent light bulb.

$$\$4.00 + \$68 = \$72$$

Chapter 7

Problem 1

STEP 1: Identify the equation to calculate heating degree days, in which HDD is heating degree days, T_{base} is the base temperature or inside temperature, usually 65°F, and T_a is the average outside temperature.

$$HDD = T_{base} - T_a$$

STEP 2: Substitute the known variables and solve for heating degree days.

$$HDD = 65°F - 13°F$$

$$HDD = 52 \text{ Degree-days}$$

Problem 3

STEP 1: Calculate the heating degree days for each day of the week.

DAY	AVERAGE TEMPERATURE	HDD
Sunday	49°F	16
Monday	47°F	18
Tuesday	51°F	14
Wednesday	60°F	5
Thursday	65°F	0
Friday	67°F	0
Saturday	58°F	7

STEP 2: To calculate the HDD for the week add the HDD for each day of the week.

$$HDD_{week} = 60 \text{ Degree-days}$$

Problem 5

STEP 1: Calculate the heating degree days for each month to determine the heating season. Then, multiply the number days in each month by its HDD to calculate the HDD for the month.

MONTH	AVERAGE TEMPERATURE	HDD	HDD_{month}
January	25°F	40	1240
February	28°F	37	1036
March	37°F	28	868
April	48°F	17	510
May	59°F	6	186
June	67°F	0	0
July	71°F	0	0
August	70°F	0	0
September	62°F	3	90
October	51°F	14	434
November	41°F	24	720
December	31°F	34	1054

STEP 2: Add the HDD for each month to calculate the HDD for the heating season.

$$HDD_{heating\ season} = 6,138 \text{ Degree-days}$$

Problem 7

STEP 1: Add the R-values of the materials the wall consists of to calculate the composite R-value of the wall.

$$Composite\ R\text{-}value = 15.2 \text{ ft}^2 \text{ °F hr/BTU}$$

Problem 9

STEP 1: Add the R-values of the materials the wall consists of to calculate the composite R-value of the wall. Remember to multiply the R-value of the fiberglass by its total thickness.

$$Composite\ R\text{-}value = 19.76 \text{ ft}^2 \text{ °F hr/BTU}$$

Problem 11

STEP 1: Add the R-values of the materials the wall consists of to calculate the composite R-value of the wall.

$$(8 \text{ in.} \times .08) + (3 \text{ in.} \times 7.00) + .94 = 22.58 \text{ ft}^2 \text{ °F hr/BTU}$$

Problem 13

STEP 1: Find the wall area of the room by adding the area for each wall.

$$Area = 16\,ft \times 8\,ft = 128\,ft^2 \times 2\,walls = 256\,ft^2$$

$$Area = 12\,ft \times 8\,ft = 96\,ft^2 \times 2\,walls = 192\,ft^2$$

$$Total\,Area = 256\,ft^2 + 128\,ft^2 = 448\,ft^2$$

Problem 15

STEP 1: Identify the equation to calculate heat loss.

$$Heat\,Loss\,(BTU/h) = \frac{Area \times (T_{inside} - T_{outside})}{R\text{-}value}$$

STEP 2: Calculate the heat loss by substituting the known variables.

$$Heat\,Loss = \frac{(10\,ft \times 8\,ft) \times (70°F - 43°F)}{22}$$

$$Heat\,Loss = 98.18\,BTU/h$$

Problem 17

STEP 1: Calculate the heat loss per hour by substituting the known variables.

$$Heat\,Loss = \frac{(360\,ft^2) \times (65°F - 37°F)}{22}$$

$$Heat\,Loss = 458.18\,BTU/h$$

STEP 2: Calculate the heat lost through the room in 12 hours.

$$\frac{458.18\,BTU}{h} \times 12\,h = 5{,}498.18\,BTU$$

Problem 19

STEP 1: Calculate the heat loss per hour by substituting the known variables.

$$Heat\,Loss = \frac{(246\,ft^2) \times [70°F - (-4°F)]}{19}$$

$$Heat\,Loss = 958.11\,BTU/h$$

STEP 2: Calculate the heat lost through the wall during a 36 hour period.

$$\frac{958.11\,BTU}{h} \times 36\,h = 34{,}491.79\,BTU$$

Problem 21

STEP 1: Calculate the seasonal heat loss by substituting the known variables.

$$Seasonal\ Heat\ Loss = \frac{(2\ ft \times 3\ ft) \times 11{,}000°F\ day \times 24\ h/day}{1}$$

$$Seasonal\ Heat\ Loss = 1{,}584{,}000\ BTU$$

Problem 23

STEP 1: Calculate the HDD for the 220 day heating season in Fargo, ND.

$$HDD = 65°F - 27°F = 38\ Degree\text{-}days$$

$$38\ Degree\text{-}days \times 220\ days = 8{,}360\ Degree\text{-}days$$

STEP 2: Calculate the seasonal heat loss through the window.

$$Seasonal\ Heat\ Loss_{window} = \frac{(8\ ft \times 6\ ft) \times 8{,}360°F\ day \times 24h/day}{1}$$

$$Seasonal\ Heat\ Loss_{window} = 9{,}630{,}720\ BTU$$

Problem 25

STEP 1: Add the R-values of the materials the wall consists of to calculate the composite R-value of the wall.

$$.81 + .94 + (3\ in. \times 3.7) + (1.5 \times 6.25) + 45 = 22.68\ ft^2\ hr/BTU$$

STEP 2: Calculate the seasonal heat loss through windows, the walls, and the roof using their respective areas and R-values.

$$Seasonal\ Heat\ Loss_{windows} = \frac{(580\ ft^2) \times 6{,}000°F\ day \times 24\ h/day}{1}$$

$$Seasonal\ Heat\ Loss_{walls} = \frac{(1{,}920\ ft^2) \times 6{,}000°F\ day \times 24\ h/day}{22.68}$$

$$Seasonal\ Heat\ Loss_{roof} = \frac{(2{,}750\ ft^2) \times 6{,}000°F\ day \times 24\ h/day}{22}$$

STEP 3: Calculate the seasonal heat lost through the house in State College, PA by totaling the seasonal heat loss values.

$$Seasonal\ Heat\ Loss_{house} = 113{,}710{,}476.2\ BTU$$

Problem 27

STEP 1: Convert CCF to BTUs of natural gas. Recall that 1 CCF is equal to 100,000 BTU.

$$34\,CCF \times \frac{100,000\,BTU}{1\,CCF} = 3.4\,MMBTU$$

Problem 29

STEP 1: Convert the seasonal heating requirement from BTUs to CCF.

$$236\,MMBTU \times \frac{1\,CCF}{100,000\,BTU} = 2,360\,CCF$$

Problem 31

STEP 1: Convert CCF to BTUs of natural gas.

$$106\,CCF \times \frac{100,000\,BTU}{1\,CCF} = 10.6\,MMBTU$$

STEP 2: Convert BTUs to kilowatt-hours of electricity.

$$10.6\,MMBTU \times \frac{1\,kWh}{3,412\,BTU} = 3,106.68\,kWh$$

Problem 33

STEP 1: Calculate the seasonal heat loss through windows, the walls, and the roof using their respective areas and R-values.

$$Seasonal\ Heat\ Loss_{windows} = \frac{(860\,ft^2) \times 9,000°F\ day \times 24\,h/day}{1}$$

$$Seasonal\ Heat\ Loss_{walls} = \frac{(2,920\,ft^2) \times 9,000°F\ day \times 24\,h/day}{19}$$

$$Seasonal\ Heat\ Loss_{roof} = \frac{(3,850\,ft^2) \times 9,000°F\ day \times 24\,h/day}{22}$$

STEP 2: Calculate the seasonal heat lost through the house by totaling the seasonal heat loss values.

$$Seasonal\ Heat\ Loss_{house} = 256,755,790\,BTU$$

STEP 3: Convert BTUs to gallons of heating oil considering the furnace efficiency is 80 percent.

$$256{,}755{,}790 \; BTU \times \frac{1 \; gal}{140{,}000 \; BTU \times 80\%} = 2{,}292.6 \; gal$$

Problem 35

STEP 1: Convert gallons to MMBTU of heating oil.

$$\frac{\$1.08}{gal} \times \frac{1 \; gal}{.14 \; MMBTU} = \frac{\$7.71}{MMBTU}$$

Problem 37

STEP 1: Convert tons to MMBTU of coal.

$$\frac{\$6.50}{ton} \times \frac{1 \; ton}{23.00 \; MMBTU} = \frac{\$0.28}{MMBTU}$$

Problem 39

STEP 1: Convert gallons to MMBTU considering the furnace efficiency is 68 percent.

$$\frac{\$1.02}{gal} \times \frac{1 \; gal}{.14 \; MMBTU \times 68\%} = \frac{\$10.71}{MMBTU}$$

Problem 41

STEP 1: Calculate the payback period by dividing the cost of the new insulation by the savings per year.

$$\frac{\$6{,}500}{\$800/yr} = 8.1 \; yrs$$

Problem 43

STEP 1: Identify the equation for estimating the cost effectiveness of adding insulation in terms of payback period, in which C_i is cost of insulation in dollars per square feet, C_e is cost of energy in dollars per BTU, E is efficiency of the heating system, R_1 is initial R-value of section, R_2 is final R-value of section, and HDD is heating degree days.

$$Payback \; period = \frac{C_i \times R_1 \times R_2 \times E}{C_e \times [R_2 - R_1] \times HDD \times 24}$$

STEP 2: Determine C_e by converting gallons to BTUs of heating oil.

$$C_e = \frac{\$1.13}{gal} \times \frac{1 \; gal}{140{,}000 \; BTU} = \$8.0714 \times 10^{-6}$$

STEP 2: Substitute the known variables and solve for payback period.

$$Payback\ period = \frac{0.60 \times 13 \times 22 \times .78}{(8.0714 \times 10^{-6}) \times 9 \times 4,600 \times 24} = 16.69\ yrs$$

Problem 45

STEP 1: Determine C_e by converting gallons to BTUs of heating oil.

$$C_e = \frac{\$1.30}{gal} \times \frac{1\ gal}{140,000\ BTU} = \$9.285 \times 10^{-6}$$

STEP 2: Determine C_i by dividing the estimated cost of instillation by the area of the windows.

$$C_i = \frac{\$8,000}{626\ ft^2} = \frac{\$12.78}{ft^2}$$

STEP 3: Substitute the known variables into the equation for estimating the cost effectiveness of installing new, energy-efficient windows.

$$Payback\ period = \frac{12.78 \times 1 \times 6 \times .80}{(9.285 \times 10^{-6}) \times 5 \times 6,500 \times 24} = 8.47\ yrs$$

Chapter 8

Problem 1

STEP 1: Identify the equation to calculate the efficiency of a heat pump, in which *COP* is the Coefficient of Performance, T_{Hot} is the heat input at a high temperature, and T_{Cold} is the heat rejected at a low temperature.

$$COP = \frac{T_{Hot}}{(T_{Hot} - T_{Cold})}$$

STEP 2: Substitute the known variables and solve for heat input.

$$12 = \frac{T_{Hot}}{(30°K)}$$

$$T_{Hot} = 360°K$$

Chapter 9

Problem 1

STEP 1: Identify the equation for calculating how many BTUs a system pulls out each hour.

$$1 \; ton = \frac{12{,}000 \; BTU}{h}$$

STEP 2: Multiply 5 tons by the conversion above to calculate how many BTUs the system pulls out each hour.

$$5 \; tons \times \frac{12{,}000 \; BTU}{h} = \frac{60{,}000 \; BTU}{h}$$

Problem 3

STEP 1: The percent reduction can be calculated using the old room air conditioner EER and the new room air conditioner EER. Divide the difference between the old EER and the new EER by the new EER. Then, multiply the ratio by 100% to calculate the percent reduction.

$$\frac{(11 - 5.5)}{11} \times 100\% = 50\%$$

Lightning Source UK Ltd.
Milton Keynes UK
UKHW050931030323
417973UK00010B/911

9 781524 924911